# Lithic Raw Material Resources and Procurement in Pre- and Protohistoric Times

## Proceedings of the 5th International Conference of the UISPP Commission on Flint Mining in Pre- and Protohistoric Times (Paris, 10–11 September 2012)

Edited by

Françoise Bostyn
François Giligny

BAR International Series 2656
2014

Published in 2016 by
BAR Publishing, Oxford

BAR International Series 2656

*Lithic Raw Material Resources and Procurement in Pre- and Protohistoric Times*

ISBN 978 1 4073 1298 9



Published with the help of Université Paris 1 Panthéon-Sorbonne/UMR Trajectoires



BAR Publishing is the trading name of British Archaeological Reports (Oxford) Ltd. British Archaeological Reports was first incorporated in 1974 to publish the BAR Series, International and British. In 1992 Hadrian Books Ltd became part of the BAR group. This volume was originally published by Archaeopress in conjunction with British Archaeological Reports (Oxford) Ltd / Hadrian Books Ltd, the Series principal publisher, in 2014. This present volume is published by BAR Publishing, 2016.

Printed in England

BAR
PUBLISHING

BAR titles are available from:

BAR Publishing
122 Banbury Rd, Oxford, OX2 7BP, UK
EMAIL info@barpublishing.com
PHONE +44 (0)1865 310431
FAX +44 (0)1865 316916
www.barpublishing.com

## Table of contents

1. Introduction.......................................................................................... 7
François GILIGNY, Françoise BOSTYN

2. Lithotheca of the Hungarian National Museum.............................................. 9
Katalin BIRO

3. Le projet LitoCat..................................................................................... 17
David ORTEGA, Xavier TERRADAS

4. A siliceous rocks database in the northwestern Paris basin : data and resources.................. 25
Claira LIETAR, Olivia DUPART, Bruno ROBERT, François GILIGNY, Françoise BOSTYN, Laurence MANOLAKAKIS

5. Travels and transformations of flint: towards a new model of 'lithotheca'.............................. 39
Paul FERNANDES, Jean-Paul RAYNAL, Anne HAUZEUR, Pascal TALLET

6. Investigation of new chipping floors in 2011 on the Spiennes Neolithic Flint Mining Site..... 49
Hélène COLLET, Michel WOODBURY, Jean-Philippe COLLIN

7. Flint mining and production of axes in the Neolithic in the Loir valley, France....................... 61
Harold LETHROSNE, Jean-Michel LECOEUVRE

8. The White Carpathian radiolarites – questions of chronology, extraction and distribution..... 73
Martin OLIVA

9. The Exploration of the Mining Site for Radiolarite in the White Carpathians Area.................. 87
Ivan CHEBEN and Michal CHEBEN

10. Features of flint mining and processing during the Chacolithic period in the Southeast of Europe (based on the materials of the Tripolian Bodaki settlement, Ukraine................................ 93
Natalia SKAKUN, Anaik SAMZUN Boryana MATEVA and Vera TEREKHINA

11. Exchange of flint materials during the Middle Neolithic – a case study from western Germany........................................................................................ 107
Kathrin NOWAK

12. Contemporary Rock Crystal Mining in Minas Gerais, Brazil – an Ethno-Archaeological Case Study........................................................................... 121
Michael BRANDL & Gerahrdt TRNKA

## Contributors

Katalin T. BIRÓ
Hungarian National Museum,
Department of Archaeology,
1088 Budapest, Múzeum krt. 14-16, Hungary
tbk@ace.hu

Michael BRANDL
Austrian Academy of Sciences,
Prehistoric Commission
Fleischmarkt 22
A-1010 Vienna Österreich
Michael.Brandl@assoc.oeaw.ac.at

Ivan CHEBEN
Archeologický ústav SAV.
Akademická 2,
949 21 Nitra. Slovakia
ivan.cheben@savba.sk

Michal CHEBEN
Archeologický ústav SAV.
Akademická 2,
949 21 Nitra. Slovakia
michal.cheben@savba.sk

Hélène COLLET
Service public de Wallonie, Service de l'Archéologie de la Direction extérieure du Hainaut I
16 place du Béguinage
B-7000 Mons Belgium
hcollet@tvcablenet.be

Jean-Philippe COLLIN
Société de Recherche préhistorique en Hainaut (SRPH)
Rue Godefroid, 2 / 21,
B-5000 Namur Belgium
collin.jeanphilippe@gmail.com

Olivia DUPART
olivdupart@free.fr

Françoise BOSTYN
Inrap Nord-Picardie & UMR 8215 Trajectoires
11 rue des Champs
59650 Villeneuve d'Ascq, France
francoise.bostyn@inrap.fr

François GILIGNY
Université Paris 1, UMR 8215, Trajectoires
Maison Archéologie Ethnologie
21 allée de l'université
92023 Nanterre Cedex, France
giligny@univ-paris1.fr

Laurence MANOLAKAKIS
CNRS, UMR 8215, Trajectoires
Maison Archéologie Ethnologie
21 allée de l'université
92023 Nanterre Cedex, France
laurence.Manolakakis@mae.cnrs.fr

Paul FERNANDES
SARL Paléotime &
UMR 5199 – PACEA,
Université Bordeaux-1, bâtiment B18,
avenue des Facultés,
33405 Talence Cedex, France
paul.fernandes@paleotime.fr

Anne HAUZEUR
SARL Paléotime,
272 rue du lycée polonais,
38250 Villard-de-Lans, France
Anne.hauzeur@paleotime.fr

Claira LIETAR
Université Paris 1, UMR 8215, Trajectoires
Maison Archéologie Ethnologie
21 allée de l'université
92023 NANTERRE Cedex, France
claira.lietar@malix.univ-paris1.fr

Harold LETHROSNE
INRAP
525 av de la pomme de pin
45590 St-CYR-EN-VAL
harold.lethrosne@inrap.fr

Jean-Michel LECOEUVRE
jm.lecoeuvre@orange.fr

Boryana MATEVA
Historical museum-Isperih,
6 Tsar Osvoboditel Str.
Isperih, 7400, Bulgaria
boryanamateva@yahoo.com

Kathrin NOWAK
PhD Student Universität zu Köln,
Institut für Ur- und Frühgeschichte
St.-Vither-Str. 1,
50933 Cologne, Germany
KaNo80@gmx.de

Martin OLIVA
Musée de Moravie
Moravské zemské muzeum, Zelný trh 6,
659 37 Brno, Rep. Czech.
moliva@mzm.cz

David ORTEGA
Spanish National Research Council (CSIC)
IMF Archaeology and Anthropology Department
C/ Egipcíaques, 15,
E-08001 Barcelona, Spain
ortega@imf.csic.es,

Jean-Paul RAYNAL
UMR 5199 – PACEA &
Departement of Human Evolution,
Max Planck Institute for Evolutionary
Anthropology,
Deutscher Platz 6,
D-04103 Leipzig, Germany
jp.raynal@pacea.u-bordeaux1.fr

Bruno ROBERT
Inrap Nord-Picardie, UMR 8215, Trajectoires
3 impasse du commandant Gérard
02260 Soissons, France
bruno.robert@inrap.fr

Anaïk SAMZUN
National Institute for Preventive Archaeology
(Inrap) & UMR 8215 Trajectoires
31 Rue de Delizy
93698 Pantin, France
anaick.samzun@inrap.fr

Natalia SKAKUN
Institute for the Material Culture History Russian
Academy of Sciences
Dvorzovaya emb., 18
St. Petersburg 191186, Russia
skakunnatalia@yandex.ru

Pascal TALLET
SARL Paléotime,
272 rue du lycée polonais,
38250 Villard-de-Lans, France

Vera TEREKHINA
Institute for the Material Culture History Russian
Academy of Sciences
Dvorzovaya emb., 18
St. Petersburg, 191186, Russia
terehinavera@mail.ru

Xavier TERRADAS
Spanish National Research Council (CSIC)
IMF Archaeology and Anthropology Department
C/ Egipcíaques, 15,
E-08001 Barcelona, Spain
terradas@imf.csic.es

Gerhardt TRNKA
Institut für Ur- & Frühgeschichte
Universität Wien
Franz-Klein-Gasse 1
1190 Wien, Austria
gerhard.trnka@univie.ac.at

Michel WOODBURY
Service public de Wallonie,
Service de l'Archéologie de la Direction extérieure
du Hainaut I
16 place du Béguinage
B-7000 Mons, Belgium

Delegates of the conference in front of the Institut d'art et d'archéologie, Paris 2012 (Photo: J. Lech).

# Introduction

Françoise Bostyn, François Giligny

The *Union Internationale des Sciences Pré- et Protohistoriques* (UISPP) commission on "Flint Mining in Pre- and Protohistoric Times" was created at the 12th meeting of the European Association of Archaeologists (Cracow, Poland, 19th-24th September 2006).
The aim was to perpetuate the tradition of organizing international symposia on flint, established by the Limburg Branch of the Dutch Geological Society in 1969 at Maastricht. The commission intends to maintain cooperation in archaeological research on siliceous rock mining (flint, chert, hornstone, radiolarite, jasper and obsidian), by presenting and discussing methods and results. Major fields of interest include the different stages of *chaînes opératoires* of manufacture, specialisation of labour and circulation of raw materials, as well as the study of flint mining sites in relation to pre- and protohistoric settlement patterns. The objective of the commission is to promote these lines of research into flint mining and its methods, thus enabling a better understanding of the various phenomena and processes taking place in pre- and protohistoric times.

Five conferences have been organised since 2007: Paris, Madrid, Vienna, Florianopolis and Paris. The Madrid conference was published in 2011 (Capote M., Consegrua S., Diaz del-Rio P., Terradas X. (eds.) *Proceedings of the 2nd International Conference of the UISPP Commission on Flint Mining in Pre- and Protohistoric Times*, BAR International Series 2260).

The fifth conference was held at Paris 1 Panthéon-Sorbonne university, on 10th-11th September 2012. The conference brought together thirty-three members of the Commission from eleven different countries. This volume contains the papers of the Paris conference, together with some additional papers presented at Vienna 2010 and Florianópolis 2011.

A first set of contributions concerns the main topic of the conference, which was *lithothèques* and reference collections.
Katalin Biro presents the Lithotheca of the Hungarian National Museum at Budapest. This contains samples collected from Hungary since 1986, as well as materials from central Europe. It was built up through systematic survey and sampling and is now published online (www.ace.hu/litot). This well known reference collection has become one of the most consulted collections of the Hungarian National Museum and is regularly updated and completed.
The Catalonian Siliceous Rock Lithotheque provides another example of reference collection documentation, management and access (Ortega & Terradas). It is relatively recent (2008) and aims to build up an exhaustive reference collection for this province. The methodology and management of the Lithotheque were particularly well designed.
A French example of a *lithothèque* is illustrated by the Soissons reference collection in the Oise/Aisne area in the Paris basin (Lietar *et al.*). This regional collection, created in the 1980s, has been updated and integrated within a database. It includes samples from the Cretaceous and the beginning of the Tertiary (Eocene and Palaeocene), in flint, chert, sandstone, limestone and other materials. It has been used for a number of regional case studies involving pre- and protohistoric archaeology.
The last contribution looks in a new way at the history of geological contexts in southern France, with particular reference to alteration processes (Fernandes *et al.*). Materials from primary and secondary deposits can thus be identified. Combined with geomorphological information, these data can be used to find outcrops.

A further group of papers concerns the second conference theme: workshops, from excavation to *chaînes opératoires* reconstruction.

The research team investigating the flint mine of Spiennes describes the new excavation carried out prior to building of a new museum (Collet *et al.*). Its construction is an outcome of the site obtaining Unesco World

Heritage status in 2005. An area with a well preserved chipping floor was excavated and part of this will be displayed in the museum. The results of the excavation are presented and the manufacturing process for flint products is analysed and reconstructed.

An overview of flint axe production and distribution in the Loir valley in France is proposed by H. Lethrosne and M. Lecoeuvre. Several flint axe production workshops and mines have been documented here since the 19th century. The *chaîne opératoire* is reconstructed, using a representative sample from different sites. This enables a broad picture to be drawn of axe production and diffusion in the area.

Two papers deal with radiolarite in the Carpathian part of Slovakia. The first, by M. Olivia, clearly outlines use of white Carpathian radiolarite from Palaeolithic to Eneolithic times. Blade technology was used during the Neolithic along the river Vlára, with evidence for workshops. This material had previously been attributed to the Palaeolithic. Then I. & M. Cheben describe the extraction pits they excavated near Horná lysá in the Chmelová Hill area of Slovakia. The radiolarite was mined by means of fire during the Neolithic.

K. Nowak investigates the middle Neolithic procurement system in the lower Rhine valley. Her work is based on previously published material. The application of factor analysis to this data set reveals two different models: intraregional exchange and extraregional procurement. Intraregional exchange involves producer-distributor settlements with higher production rates.

The last papers are *miscellanea.* M. Brand and G. Trnka describe a present-day rock crystal mine in Brazil, exploited by individual, independent miners with the permission of the land owners. It provides an interesting analogy for the extraction conditions and social context of flint mining in Neolithic Europe.

# Lithotheca of the Hungarian National Museum

Katalin T. Biró


## Abstract

The comparative raw material collection of the Hungarian National Museum was founded in 1986, as a result of:
(1) collecting activity of the Hungarian Geological Institute from the late seventies,
(2) an international conference organised by the HGI and the HNM on the subject of flint mining and lithic raw material identification in the Carpathian Basin (Biró ed. 1986, 1987).
The starting stock involved a systematical survey of Hungarian chipped stone raw materials, complemented by donations of colleagues from various countries in, mainly, Central Europe. After this time, the development of the collection followed normal museum procedures; special collecting trips, exchange, donations. Role of thematical projects (for chipped stone raw material research and historic quarries) extended the scope from merely chipped stone tools to general lithic raw materials used in prehistory.
Catalogues of the collection were published so far in two volumes (Biró & Dobosi 1991, Biró *et al.* 2000); currently, we are working on the third volume that will be published in electronic format.


**Keywords**: comparative raw material collection; Lithotheca; Hungarian National Museum.

## 1. Introduction

The Lithotheca of the Hungarian National Museum is one of the regular museum collections of the museum. It is operating as part of the Prehistory Department, notably, the Palaeolithic Collection. Its goal is to provide reference material for characterisation studies of archaeological lithic assemblages. The original scope centred on, lithologically, chipped stone tool raw materials, mainly siliceous rocks and obsidian, and, from a geographical point of view, the Carpathian Basin. In the course of time, however, the scope was extended to other rocks used by prehistoric peoples for the production of polished stone tools and other stone utensils. Geographically, we are still concentrating on Central Europe Middle Danube and the Basin but occasional collecting trips and much appreciated donations and exchange widened the geographical horizon to five continents.

The collection serves for supporting petro-archaeological analysis and characterisation studies, basically for prehistoric people living on the territory of present-day Hungary. By the parallel study of archaeological lithic assemblages and the source areas we believe to have a good coverage from Hungary in general; however, to solve basic tasks of petroarchaeological characterisation, at least the area of the Carpathian Basin should be fully represented. Another problem is the accessibility of the former sources; some were fully exploited, others were uncovered/hidden by changing geological conditions. Looking back to field experiences over a few decades we can observe that even during one or two generations, sources may disappear as well or become very difficult to access.

## Origins

The Lithotheca of the HNM was founded in 1986 as a result of several years of fieldwork mainly performed at the Hungarian Geological Institute under the auspices of József Fülöp. The Hungarian Geological Institute was implemental in early petroarchaeological characterisation studies, starting from the first Hungarian Palaeolithic sites (Kadić 1915, Vendl 1930 etc.) and, later on, took important initiative in the excavation and preservation of flint mines (Fülöp 1973, 1976, Bácskay 1980, 1990). József Fülöp coordinated a general research program on mineral raw materials used in prehistory as well as historical periods (Fülöp 1984). His monograph has remained a milestone in prehistoric / historical raw material research, complemented by modern studies very recently (Szakáll 2008).
Systematical collection of lithic raw materials was started, partly, in connection with the characterisation effort of flint mines (Sümeg and Tata especially) and tracing their products on Hungarian prehistoric sites. The other important framework was the documentation

of lithological formations and key sections (Császár 1997) for the collection of the HGI. By 1986, the level of knowledge on, at least, the chipped stone raw materials in Hungary allowed the organisation of a reference collection (Biró 1984, Biró & Pálosi 1986). The international aspect was supported by the organisation of a conference in Sümeg, at the open-air museum established near the largest flint mine known from Hungary up to present days (Biró 1986, 1987). The conference had wide international participation, and participants were specially asked to contribute to a basic reference collection, to be installed at HNM, by some of the most characteristic chipped stone raw material types in their spheres of interest and activity. The systematical field surveys in the early 1980's and the donations of the Sümeg Conference participants constitute the basic fond of the Lithotheca Collection.

## 2. Acquisition

Since 1986, the Lithotheca is available for researchers at the Hungarian National Museum. It is treated exactly like the other collections of the Museum; it has curatorial care, inventory system according to general museum regulations in Hungary and a more or less systematical acquisition strategy. The Lithotheca is constantly increasing: see fig.1. for details. The "gaps" in the graph show periods of intensive work: compilation of the catalogues (Biró-Dobosi 1991, Biró *et al.* 2000) and conversion to official computerised inventory of the HNM (MUNYIR, Biró 2008).
The acquisition strategy involves more or less regular collecting trips, within and outside Hungary, often supported by national, bilateral or international grants. Some of the more important projects include the so-called Upper Tisza project lead by J. Chapman (Chapman 1994), when K.T. Biró and K. Hardy systematically surveyed all mapped siliceous rock deposits of the Tokaj Mts. (Hardy & Biró 1993), national scientific foundation (OTKA) projects, notably OTKA T-0136638, "*Archaeometrical study of stone artefacts, 1994-1997*", and T-025068, *Atlas of non-metallic raw materials, 1998- 2002"*, implemental in collecting polished stone tool raw materials (Biró 1998, Biró et al. 2004, http://www.ace.hu/atlas/), The IGCP-442 project, supported by UNESCO, also resulted in the accumulation of evidence on polished stone raw materials (Hovorka & Meres eds., http://www.ace.hu/igcp442/index.html). *Erosion and humidity*, later on *Historic quarries* projects were contributing most to the research of quernstones and polishers (Balak 2005, http://chc.sbg.ac.at/saxahq/). Among the bilateral programs, the Greek-Hungarian collaboration project (TÉT 1997-1999, Biró and Dobosi 1998) has proved to be invaluable for obsidian research whereas the Croatian-Hungarian collaboration program (TÉT 2008-2009, Težak-Gregl & Burić 2009) helped to fill the gap in the flow of petroarchaeological information to the south of our borders. Collaboration with Tübingen University contributed to the Lithotheca by marble source reference samples from our region (MÖB-DAAD project 2001-2002, Zöldföldi *et al.* 2005).
The Lithotheca can always count on the donations of colleagues, on fieldwork or just vacation abroad. They do not always bring material of primary scientific interest; nevertheless they are important and integrated parts of our collection. We operate regular exchange service, offering authentic samples from Hungarian source collected lithic raw materials and receive valuable specimens in return. Some of the more important sources of exchange involve colleagues like Elisabetta Starnini, Otis Crandell, Tincu Sorin etc.
Last but not least, the Lithotheca is constantly increased by control material of petroarchaeological analyses on archaeological finds and evidences of experimental work on various lithic raw material sources (replicas produced by, e.g., J. Weiner, K. Simán, Z.H.Tóth).

## 3. Coverage

As it was stated before, the main purpose of the Lithotheca of the HNM is to serve as reference collection for Hungarian prehistoric localities. This goal determines our focus of interest. However, it is well know, that some raw materials of high quality and prestige travelled in the range of several hundred, sometimes over thousand kilometres already in prehistoric times. To identify and characterise these materials is one of the most important and challenging tasks of petroarchaeology. It is imperative to collect, stock and analyse as much of these *Long Distance* and *Extra-Long Distance* materials as we can, because they are at least as important for modern archaeological research as they were for our picky ancestors.
In the case of basically local use of, mainly, household stone utensils – quernstones, hammerstones, polishers – we are facing another type of problem; that is, quite a lot of „stones" were suitable and, in fact used, for basic tasks, leaving us with infinite possibilities to collect as reference (Fig. 2).

## 4. Publication

The first catalogue of the collection was published in 1991 (Biró-Dobosi 1991), followed by a second volume in 2000 (Biró *et al.* 2000). From the beginning, the collection is supported by a database that was integrated into the official inventory database system of the HNM. Internet publication started with the base Hungarian fond (www.ace.hu/litot)
Currently we are planning to:
(1) present existing catalogues in electronic form,
(2) organise Vol. 3. of the catalogue (probably in electronic format),
(3) complete data service with full collection and (much) better images.

## 5. Database history

The history of the HNM Lithotheca is also a history of databases. We were always trying to organise the collection data into n easily accessible, retrievable form with facilities for import and public access (Fig. 3.). The database had to fullfil various tasks, as possibilities allowed: information storage/retrieval at the basis, complemented by systematical/unified description, vernacular / English content, multi-session access, image storage, inventory purpose and internet access. Unfortunately, with every new version, apart from obvious advantages, we were loosing something. Currently, we are facing an institution-wide data migration of inventory databases effecting also the Lithotheca collection.
The original version of the Lithotheca database was realised in DBase (1987-1990), in the form of individual tables for inventory data (in Hungarian), its mirror translation (in English), register of instrumental analytical work as well as archaeological data recording (Biró 1990). The whole system was efficiently connected in a relational database, DataEase (1991-2002), which is partly still used (remodelled Biró 2005), due to the excellent relational capabilities of the program used. The DataEase version served as the basis of both catalogue volumes (Biró-Dobosi 1991, Biró *et al.* 2000) and the internet data on the basic fond of the collection (1986, hu, en http://www.ace.hu/litot). For the possibility to store images within the system, we had a short fling with MS Access (~1998), supplied with user's menu and (small) images. The installation of HNM official inventory database MUNYIR (2002-2012), an application written for Oracle database systems seemed to solve several problems, i.e., multi-session functions, attachment of images and special documents (e.g. analytical results) and potential web export (Biró 2008). The price we had to pay, in respect of information, was a strict restriction on contents (having to accommodate the very specific Lithotheca data content to general inventory rules) and loosing the bilingual facilities. To put it simple, what is good for „everybody" will not exactly fit the specialists. For this reason we are experimenting with a new structure/concept, successfully applied at various archaeometry databases (Zöldföldi *et al.* 2008, 2010). The programming framework is MS SQL. Most recently, an institutional decision was made to change from the MUNYIR system to another inventory system for HNM, called Hunteka. We have no experience with this system as yet but will try to migrate data with smallest possible loss in data content

## 6. Analytical work

The Lithotheca offers help in selecting "best practice" for valuable archaeological material. From the very first steps we are trying to examine the items of our reference collection with various techniques in view of fingerprinting the sources, checking and enhancing the effectiveness of macroscopical identification of raw material types. In Vol.1. of the Lithotheca we could enumerate 1187 analytical studies; in Vol. 2., altogether 1790. Since 2000, the number of analytical studies at least doubled due to the development in non-destructive techniques (Biró *et al.* 2002, Kasztovszky *et al.* 2008) and activity of collaboration work (Zöldföldi et al. 2005, Halamić & Šošić-Klindžić 2009, Perhoč 2009 etc.) and students' contributions (Füri 2004, Judik 2001, 2009, Péterdi 2005 etc.)

## Conclusions and the way ahead

The Lithotheca is one of the most popular, most visited and dynamically developing collections of our Museum. We are trying to ensure the information value of the collection by regular surveys and analytical work. In this spirit we are planning to edit the 3rd volume of the Catalogue in the near future, in electronic format, with better and more complete illustrations and more precise locations for the sources, within reasonable limits.
Although the story of the Lithotheca of the HNM is basically a success story so far, we have to stop at this point to make some considerations. This is necessary from both the practical and the theoretical point of view.

First of all, about the limits and future directions of developing the collection. In a philosophical sense, a collection can never be complete, there are new and new elements emerging but there must be a technical limit. The purpose of the Lithotheca is to serve regional petroarchaeological studies and it is justified as much and as long as it is serving this purpose. Obviously, at more distant territories, the relevance of our Lithotheca will be less important and we have no control on the reliability and coverage of data. By emerging regional comparative raw material collections this problem may have a solution as we have seen in collecting such information (Biró 2011). At the same time, most of such collections seem to have an ephemeral character – existing so far as the interest of the specialist working on them lasts. In the course of our work we have also encountered sad experiences on important thematical collections disappearing and forgotten in course of time – it is enough to mention the Schafarzik collection of decorative- and building stones (Schafarzik 1904) or another important thematical collection also of primary interest to archaeology, compiled on viable potters' clay deposits by S. Kalecsinszky (1905). The museum background seems to offer certain protection but we know, concepts may change and it is only the consensus of the profession that may vindicate a *raison d'être* for a comparative prehistoric raw material collection, i.e. Lithotheca.

**Acknowledgements**

The author is grateful for the current OTKA project K-100385, *„Provenance study of lithic raw materials of stone tools found in the Carpathian Basin"* lead by Zsolt Kasztovszky, facilitating participation on the current U.I.S.P.P. meeting and also supplying possibility to further develop the knowledge basis behind the Lithotheca of the HNM.

## References

Bácskay, E. 1980. Zum Stand der Erforschung prähistorischer Feuersteinbergbau. In Weisgerber, G. (ed.), *5000 Jahre Feuersteinbergbau.* Deutschen Bergbau-Museum Bochum, 179-182.

Bácskay, E. 1990. *La mineria prehistorica en Hungria.* Cuadernos de Prehistoria de la Universidad de Granada 1986 (11), 273–325.

Balak, M. 2005. Erosion and Humidity: a Culture 2000 project for the study of building materials by scientific methods. An interim report. *Archeometriai Műhely / Archaeometry Workshop Budapest 2 / 1*, 60-61.

Biró, K. T. 1984. *Őskőkori és őskori pattintott kőeszközeink nyersanyagának forrásai / Sources of palaeolithic andprehistoric stone tools.* Archaeológiai Értesítő, Budapest, 111 42–52.

Biró, K. T. 1990. *A micro-computer database system for the investigation of lithics.* In: Voorips, A. and Ottaway, B. eds., New Tools from Mathematical Archaeology. Kraków, 107-114.

Biró, K. T. 1998. The study of polished stone implements in the Carpathian Basin. In Költő, L. and Bartosiewicz, L. eds., *Archaeometrical Research in Hungary II.* Budapest / Kaposvár Hungarian National Museum / Directorate of Somogy County Museum, 115-139.

Biró, K. T. 2005. Gyűjtemény és adatbázis: eszközök a pattintott kőeszköz nyersanyag azonosítás szolgálatában / Collection-and-Database Approach in the Study of Lithic Raw Material Provenance *Archeometriai Műhely / Archaeometry Workshop Budapest 2005 2 / 4*, 46-51.

Biró, K. T. 2008. A múzeumi nyilvántartás számítógépes rendszerének hivatalos bevezetése - lehetőségek, eredmények, problémák. Gondolatok az első sikeres auditok kapcsán / The official introduction of computerised system in museum documentation - potentials, achievements. *Múzeumi Közlemények Budapest 2008/2*, 54-65.

Biró, K. T. 2011. Comparative raw material collections in support of petroarchaeological studies: an overview / Összehasonlító nyersanyaggyűjtemények a petroarcheológiai vizsgálatok szolgálatában: áttekintés. In Biró, K.T. and Markó, A. eds., *Emlékkönyv Violának.* Budapest, Magyar Nemzeti Múzeum, 225-244.

Biró, K. T. and Pálosi, M. 1986. A pattintott kőeszközök nyersanyagának forrásai Magyarországon / Sources of chipped stone raw materials in Hungary. *Magyar Állami Földtani Intézet Évi Jelentése*, Budapest, 407-435.

Biró, K.T. and Dobosi V., 1991. *LITOTHECA - Comparative Raw Material Collection of the Hungarian National Museum.* Budapest, Magyar Nemzeti Múzeum, 1-268.

Biró , K.T. and Dobosi, V.T. 1998. Görög-magyar petroarcheológiai együttműködési program. / Greek-Hungarian petroarchaeological collaboration program. *IRAMTO* XV-XVI http://www.ace.hu/iramto/98gorogmagyar.html

Biró, K. T., Dobosi, V. T. and Schléder, Zs., 2000. LITOTHECA - Comparative Raw Material Collection of the Hungarian National Museum. Vol. II. – Budapest, Magyar Nemzeti Múzeum, 1-320.

Biró, K. T., Elekes, Z., Uzonyi, I. and Kiss, Á. 2002. Radiolarit minták vizsgálata ionnyaláb analitikai módszerekkel / Investigation of Radiolarite Samples by Ion-Beam Analytical Methods. *Archaeológiai Értesítő 127*, 103-134.

Biró, K. T., Scharek, P. and Szakmány, Gy. 2004. Towards an atlas of prehistoric (non-metallic) raw materials in the Carpathian Basin. In *Actes du XIV Congres UISPP Section 2. Archeometry*. BAR International Series 1270, Oxford, 57-60.

Biró, K.T. (ed), 1986. *Őskori kovabányászat és kőeszköz-nyersanyag azonositás a Kárpát medencében / Proceedings of the 1st International conference on prehistoric flint mining and lithic raw material identification in the Carpathian Basin*. Budapest-Sümeg, Vol.1, 1-342.

Biró, K.T. (ed), 1987. *Őskori kovabányászat és kőeszköz-nyersanyag azonositás a Kárpát medencében / Proceedings of the 1st International conference on prehistoric flint mining and lithic raw material identification in the Carpathian Basin*. Budapest-Sümeg, Vol.2, 1-284.

Chapman, J. 1994. Social power in the early farming communities of Eastern Hungary - Perspectives from the Upper Tisza region. *A Jósa András Múzeum Évkönyve Nyíregyháza*, 36, 79-91.

Császár G. ed. 1997. *Magyarország litosztratigráfiai alapegységei.* MÁFI-Magyar Rétegtani Bizottság, Budapest. 114 p.

Fülöp, J. 1973. Funde des prähistorischen Silexgrubenbaues am Kálvária-Hügel von Tata. *Acta Archaeologica Academiae Scientiarum Hungaricae* 25, 3–25.

Fülöp, J. 1976. Relics of prehistoric flint mining in Hungary. In Engelen, F.G. H. (ed.), *Proceedings of the Second International Symposium on flint*, Staringia 3, 72–77.

Fülöp, J. 1984. *Az ásványi nyersanyagok története Magyarországon*. Budapest, Műszaki Könyvkiadó, 179 p.

Füri, J., Szakmány, Gy., Kasztovszky, Zs. and Biró, K.T. 2004. The origin of the raw material of basalt polished stone tools in Hungary *Slovak Geological Magazine Bratislava 10 / 1-2*, 97-104.

Halamić, J. and Šošić-Klindžić, R. 2009. Radiolarites and radiolarian cherts in Northern Croatia - possible sources for the production of artifacts / Radiolarit és radioláriás tűzkő Észak-Horvátországban - lehetséges kőeszköz nyersanyagforrások. *Archeometriai Műhely / Archaeometry Workshop Budapest 6 / 3*, 19-24.

Hardy, K. and Biró, K.T. 1993. Raw material sourcing in the Tokaj mountains (manuscript field notes) Edinborough-Budapest.

Hovorka, D. and Meres, S., eds. 2001. IGCP/UNESCO project 442: "Raw materials of the Neolithic/Aeneolithic polished stone artefacts: their migration paths in Europe" *Slovak Geological Magazine, Bratislava Geological Survey of Slovak Republic 7 /4*, 313-436.

Judik, K., Biró, K. T. and Szakmány, Gy. 2001. Petroarchaeological research on the Lengyel Culture polished stone axes from Aszód, Papi földek. In: Regenye, J. (ed.): *Sites and Stones: Lengyel culture in Western Hungary and beyond.* Veszprém, 119-129.

Judik, K., Dobosi, G., Markó, A., Biró, K.T. and Szakmány, Gy. 2009. Electron microprobe analysis as a useful tool for the provenance characterisation of polished stone artefacts – case study from N Hungary. In *Proceedings ISA2006*, CELAT, Université Laval Québec 173-178.

Kadić O. 1915. A Szeleta-Barlang kutatásának eredményei. *Magyar Állami Földtani Intézet Évkönyve* 23, 147–278.

Kalecsinszky, S. 1905. *A magyar korona országainak megvizsgált agyagai.* [Clay sources analysed on the territory of the lands belonging under the Hungarian Holy Crown] Magyar Kir. Földtani Intézet, Budapest, p. 218.

Kasztovszky Zs., Biró K. T., Markó A. and Dobosi, V. 2008. Cold Neutron Prompt Gamma Activation Analysis—a Non-Destructive Method for Characterization of High Silica Content Chipped Stone Tools and Raw Materials. *Archaeometry 50 / 1*, 12-29.

Perhoč, Z. 2009. Sources of chert for prehistoric lithic industries in Middle Dalmatia / Adatok Közép-Dalmácia őskori kőeszközeinek nyersanyag-forrásaihoz: kovakőzetek *Aműhely Archeometriai Műhely / Archaeometry Workshop Budapest 6 / 3*, 45-56.

Péterdi, B., Kovács, T., Szakmány, Gy. and Biró, K.T. 2005. Petrographic investigation of Bronze and Iron Age Casting Moulds from the Collection of the Hungarian National Museum. In: Kars, H. and Burke, N. eds., Proceedings of the 33rd International Symposium on Archaeometry, 22-26 April 2112, Amsterdam. *Geoarchaeological and Bioarchaeological Studies* 3, 87-90.

Schafarzik F. 1904. *A magyar szent korona országainak területén létező kőbányák* [Stone quarries on the territory of the lands belonging under the Hungarian Holy Crown], Budapest.

Szakáll S. (ed.) 2008. Alkalmazott ásvány- és kőzettan. Az ásványok és az ember a mai Magyarország területén a XVIII. század végéig. Bányászat Geotudományok. A Miskolci Egyetem Közleménye. *A sorozat Miskolc Egyetemi Kiadó* 74, 1–253.

Vendl A. 1930. A büdöspesti paleolitos szilánkok közettani vizsgálata *Matematikai s Természettudományi Értesítő* 47, 468–483.

Težak-Gregl, T. and Burić, M. 2009. Archaeometrical research of lithic raw materials for early neolithic prehistoric communities with the help of Prompt Gamma Activation Analysis: the aims of project, current achievements and future perspectives. *Archeometriai Műhely / Archaeometry Workshop Budapest 6 / 3*, 1-4.

Zöldföldi, J., Taubald, H., Pintér, F., Tóth, M., Biró, K.T., Satir, M., Mráv, Zs., Kasztovszky, Zs., Szakmány, Gy. andDemény, A. 2005. Provenance studies on Roman Marble Fragments in the Hungarian National Museum. In: KARS--BURKE eds. 2005, Proceedings of the 33rd International Symposium on Archaeometry, 22-26 April 2112, Amsterdam *Geoarchaeological and Bioarchaeological Studies* 3, 119-122.

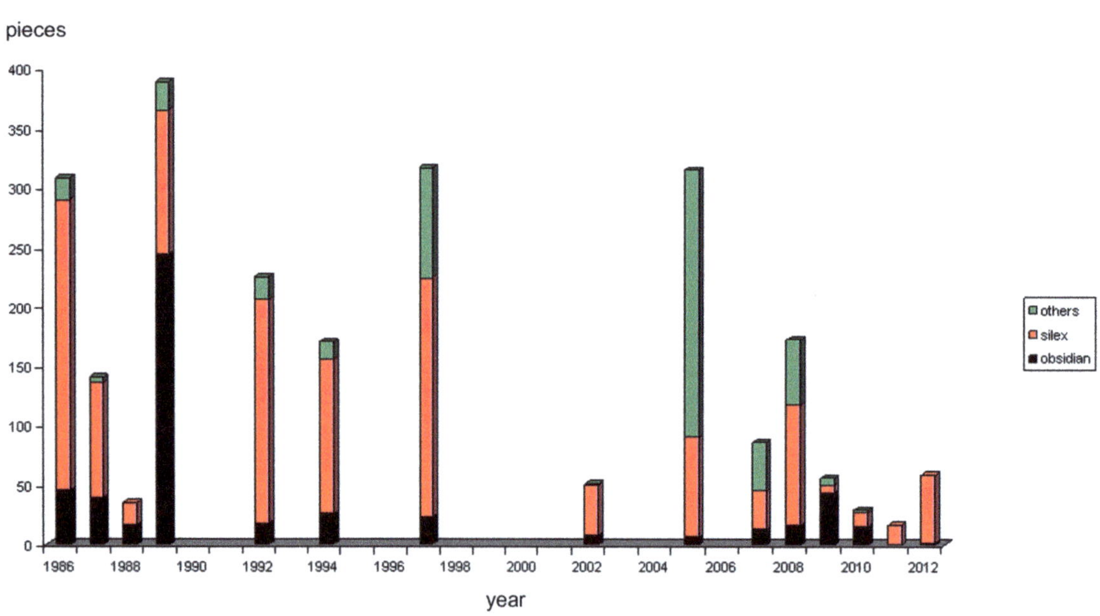


Fig. 1 : Acquisitions of the Lithotheca of the HNM by year and raw material types.

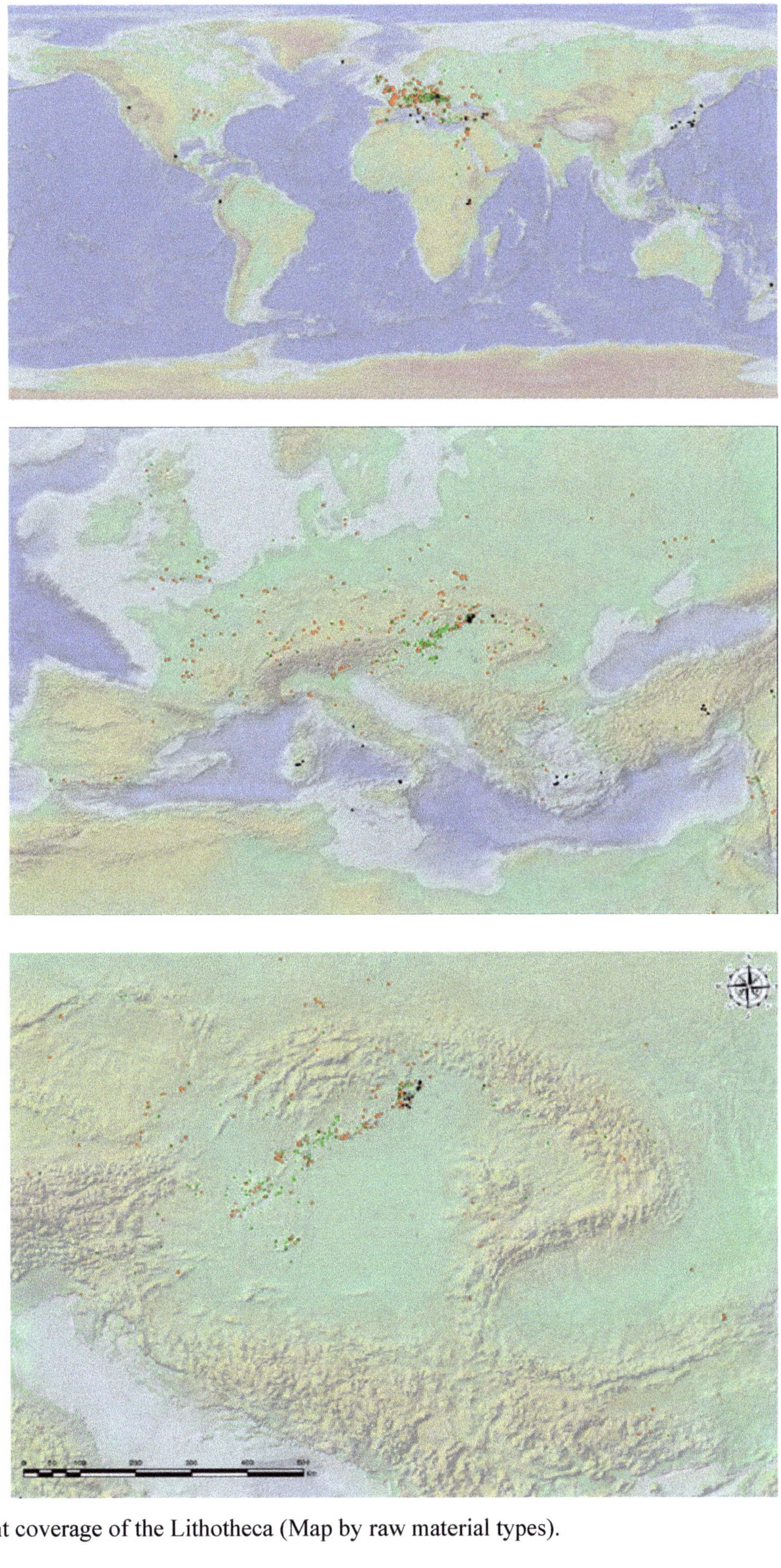

Fig. 2 : Current coverage of the Lithotheca (Map by raw material types).

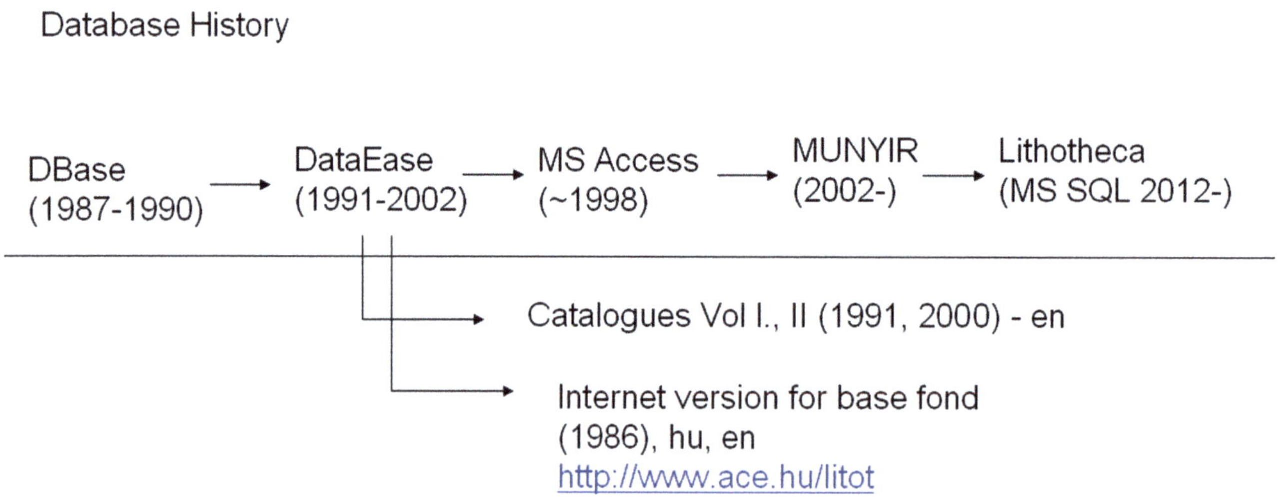


Fig. 3 : Timeline flow of Lithotheca databases.

Zöldföldi, J., Hegedüs, P. Székely, B. 2008. MissMarble - egy archeometriai, művészettörténeti és műemlékvédelmi célú, internet-alapú, interdiszciplináris adatbázis / Interdisciplinary data base of marble for archaeometric, art historian and restoration use *Archeometriai Műhely / Archaeometry Workshop Budapest 5 / 3*, 41-50.

Zöldföldi, J., Leno, V., Székely, B. Szilágyi, V., Biró, K.T. and Hegedűs P. 2010. CeraMis: interactive internet-based information system on Neolithic pottery. *Archeologia e Calcolatori* 21, 301-314.

# The lithotheca of siliceous rocks from Catalonia

David ORTEGA & Xavier TERRADAS


## Abstract

The main objective of this research project is to provide a research platform open to the scientific community interested in the availability of siliceous rocks during Prehistory in the north-eastern Iberian Peninsula and adjacent areas. With this aim in view, one of the first steps has been to create a reference collection of siliceous rocks intentionally planned to transcend the common model of lithotheques and provide the means to serve and be of use to local and outside researchers. Work has therefore focused on all the resources that siliceous rocks may provide, whether or not they were exploited in prehistoric times.

In order to achieve this, it was first necessary to build up and document collections, which should be thorough, complete and representative of different types of siliceous rocks in the geographical and geological entities included in the project. However, the collections are not in themselves a lithotheque; they must be documented, and samples have to characterized and installed so they can be accessible to users in different ways. We thus have to handle a large amount of documentary evidence in different formats (map references, photographs, laboratory reports...). It is therefore obvious that a solid infrastructure is required, as well as equipment and scientific and technical staff to manage it. In our opinion all these steps must be involved in the creation of a lithotheque.


**Keywords :** Siliceous rocks; Lithotheque; Reference collection; north-eastern Iberia; Prehistory; LitoCAT collection.

## 1. Introduction

Since Archaeology's beginnings as a scientific discipline, the formation of lithotheques and, in general, reference collections of different kinds of materials, has been a resource commonly used by archaeologists in many of their research programmes.

"Lithotheque" is the term widely used for rock and mineral reference collections. They are not only used by prehistorians, although they may be the collective that has developed them most in recent decades. Lithotheques are also used, for example by classical and medieval archaeologists interested in identifying rock types and their sources. An example is the marble that in Antiquity was widely used in architecture, sculpture and funerary monuments, and was the object of international trade (Borghini 1989, Herz and Waelkens 1988). For the same reasons, public and private architects, restorers, etc. who work in the management and conservation of the enormous stone-built heritage in old European cities also need to consult lithotheques (see Prikryl and Török 2010). Apart from archaeologists, many other agents create, use and maintain lithotheques as part of their normal duties, for different purposes. They include state organisations and institutions with public responsibility in such fields as geology, mining and map-making (e.g. the Instituto Geológico y Minero –IGME- in Spain, and the Bureau de Recherches Géologiques et Minières –BRGM- in France, to give just two examples). Universities, science museums and different kinds of research centres working not only in archaeology, may use lithotheques, while in the private field, collectors, commercial suppliers of stones used for everything from jewellery to building materials, and certainly mining and oil-drilling companies also hold rock collections.

Lithotheques are therefore more common in our society than might be thought when first approaching the subject. Without any exact data, we can safely state that the number of collections and the agents and actors creating them, managing them or simply consulting them are steadily increasing. In this way, the number of experiences involving lithotheques and their familiarity to more diverse social collectives is also increasing.

Like any other infrastructure and scientific equipment, the cost-effectiveness of reference collections ultimately depends on the use made of them by researchers and the other agents who consult those (Suárez and Tsutsui, 2004, about biological collections). However, many archaeological lithotheques were not originally formed to provide a service to any other users apart from the researchers or groups that founded them (see Biró 2011). This is usually because they were formed as a resource for research programmes that needed a collection of local materials with which the archaeological objects being studied could be compared (Turq 2005).

A lithotheque is therefore not an end in itself, but only the means to reach the objectives set by the research. As a result, many lithotheques hold collections with incoherences in the materials they contain. This is normally because of failings in the planning and definition of the thematic and geographical boundaries of their specialisation. In turn, this is usually the reason for the diversity in the lithologies selected and the multiple areas of provenance, which are often distant and lacking any territorial continuity.

Of all the lithotheques formed for archaeological research, few have been planned with a view to gathering systematically materials representative of a coherent geographic or geological unit. Probably the best example of a well-planned collection is the Hungarian National Museum Lithotheque, in Budapest, with siliceous rocks from all over the country and which regularly publishes its catalogue and description (Biró and Dobosi 1991, Biró *et al.* 2000).

## 2. The Catalonian Siliceous Rock Lithotheque

We have been working on the formation of a siliceous rock lithotheque in Catalonia, north-eastern Iberia, since 2008. Going beyond the most usual model of a lithotheque, we aim to build up a complete reference collection with all the siliceous rock types in this part of Iberia. More importantly, the project is expressly planned to provide a service for the whole scientific community interested in the subject (Terradas *et al.* 2012).

To achieve the first goal, the priority is to gather and document the collections, which must be exhaustive, complete and representative of the different kinds of siliceous rocks in north-eastern Iberia. For this, appropriate infrastructures and equipment are needed, as well as the technical and scientific staff that can take charge of the project and manage its scientific achievements. The foundation of such a lithotheque is a complex and costly process. It is therefore necessary to plan the actions to be carried out in detail and establish a calendar for their execution (see Terradas *et al.*, 2012 for further details).

The creation of a public service with this new resource was an ineluctable objective and, without doubt, the most ambitious aspect of the project. To be able to provide a valuable service, as well as constructing and storing the collection, a catalogue and full description must be published, in addition to the rules and regulations for consulting and using it.

From the start, all the processes and stages in the formation of the lithotheque have been documented. At the same time as this is a very advisable management technique in general, it should contribute towards favouring the dissemination of the experience.

Antecedents and Objectives

The precedent of the current project was a thematic network of scientific collaboration in 2003 and 2004, in which eight different groups linked to international universities and research centres took part. The two main objectives of this network were (Terradas *et al.* 2006):

- To pool the data obtained until that time by the various groups, and explore the potential of joint data processing;
- To establish common methodological procedures both in fieldwork (surveying, documentation of outcrops, etc.) and in the laboratory (archaeometric characterisation of the samples).

Despite achieving some partial results, this initiative did not fulfil all the scientific expectations. The project was then completely re-designed by the team that led the initiative (CSIC-IMF, Barcelona), on a new organisational and institutional basis which has led to the current project.

As stated above, the ultimate objective of the lithotheque is to provide a service to the users who might consult it. For this it is necessary to:

1. Create the collections, i.e. gather the samples, catalogue and document them, and store them appropriately in optimal conditions for their conservation.
2. Set up the collections and equip the spaces of the lithotheque. For this, it was necessary for the CSIC to provide rooms for stores and areas for the users (laboratories and consultation rooms). These had to be refurbished with the facilities and necessary instruments and equipment to access the collections and work with them.
3. Implement the material and organisational means enabling the effective consultation of the collections. This included, among other aspects, providing a catalogue of the services and rules for their use, digitising all the information regarding the collections and creating databases with all these data, in order to make their inventory available.

As the lithotheque is intended to be a reference collection, to maximise its usefulness and scientific value, the following requirements also needed to be fulfilled:

- Representativeness: the collections should be representative of all the siliceous rocks in Catalonia and of all their variability. To satisfy this condition, the process of acquiring the samples should be carried out systematically, according to clearly defined criteria applied rigorously, to make the collections exhaustive (they contain every type) and complete (none are missing).
- Contextualisation: all the materials should be acquired together with documentation ensuring their correct geographic and geological context;
- Accessibility: access to the collections should be guaranteed for all the collectives (researchers, academics, teachers and students) who are their potential users, and free scientific use should be encouraged in accordance with the objectives specified above.

Collections

The Catalonian Siliceous Rock Lithotheque holds the samples forming the LitoCAT collection, with siliceous rock specimens from Catalonia and neighbouring regions over a large circum-Pyrenean area covering about 50.000 km$^2$ (Fig. 1).

From the geological point of view, this area includes three main morpho-structural zones that are related in their origin and evolution: 1) the Pyrenees, the narrow high Alpine range on the union between the Iberian micro-plate and the European plate; 2) the Ebro basin, a large Tertiary sedimentary basin to the south of the Pyrenees; and 3) the Iberian System, including the Catalonian coastal ranges, which constitutes the largest of the Iberian intra-plate structures that absorbed the huge tectonic tension caused by the collision of Iberia with Europe, beginning in the late Mesozoic and continuing active until the early Neogene. This large region with such a diverse geological substrate contains a wide range of siliceous rocks of very different ages, petrogenesis and characteristics.

The collection was mainly conceived for scientific use. Its normal users are young researchers studying for their doctoral theses, consolidated research teams working in projects on prehistoric sites in the region, and foreign teams interested in particular Catalonian lithologies. The samples have been acquired in the field by part of the technical and scientific team working in the lithotheque, and to satisfy the users' demands they are documented fully and in detail. External laboratories are usually asked to prepare different types of samples, like thin sections and polished plaques, which can be consulted and studied using the lithotheque equipment, as well as some specific tests, such as crystallography, magnetic parameters and geochemical characterisation of the main and trace elements. Due to the interdisciplinary character of the institution hosting the lithotheque, most of these tests are carried out at other CSIC centres located in Barcelona.

## 3. Working methods and process

Although the project had started earlier, the work programme was re-defined in 2008 in order to achieve the objectives that had been set in the middle-term, within five years. This programme included several tasks: 1) definition of objectives and working plan; 2) provision of rooms, furniture and equipment; 3) documentary research; 4) fieldwork and acquisition of samples; 5) inventory and installation of the collections; 6) documentation and characterisation of the samples; 7) dissemination of the contents on Internet and 8) user service.

Each of these tasks involves its own methodology and techniques, which are sometimes elementary and other times quite sophisticated, and occasionally technicians and outside services need to be recruited (computer specialists, analysts, etc.). However, the planning of the different activities is not a simple sequential order of tasks to be done (Fig. 2). Many of them are often linked together to achieve a common objective and it is therefore necessary to design the stages and plan them well beforehand.

Three of these stages will be described in greater detail: the criteria for the choice of samples, their inventory, and the management of the information documenting them.

Sample acquisition

In order to build up a reference collection with well-defined rocks and avoid identification errors, fieldwork has aimed exclusively to locate and sample outcrops in a primary position. We are aware of the importance of outcrops in a secondary position in the procurement of different raw materials throughout Prehistory, but we think that this method ensures the optimal identification of the rock type and stratigraphic contextualisation. Whenever possible, the rocks have been obtained at locations described in previous geological studies. In this way the samples benefit from a large amount of information about the context, such as the lithology of the host-rock, sedimentary environment when the rock formed, micro-palaeontological record, age, and formation processes. This enables enhanced interpretation of the origin of the rocks, as well as the identification of distinctive characters within the regional framework. With this kind of sampling, the quantity and quality of the information documenting the collection's samples is increased, optimising the available data and making the work carried out more effective.

The selected samples are representative of the different kinds of siliceous rocks and their respective host-rock. They are all referenced in the corresponding record card for the outcrop, as well as by schematic diagrams drawn in the field or directly in geological documents and publications. All the samples are identified by a unique code, their stratigraphic location is identified with geographical coordinates, and they were photographed in their original position, indicating their orientation. In all cases, an additional number of samples is collected but not referenced, and care is taken to include enough fragments to cover the intrinsic variability of formats, modules and alterations of each outcrop.

Sample Cataloguing

After the fieldwork, the samples, field-notes and photographs are deposited in the lithotheque, where they are catalogued and stored. The samples are entered in the project databases using a computer application that automatically identifies them with a correlative registry number which from that point onwards becomes their sole valid reference.

The inventory is not used to describe the samples. Its only function is to manage the lithotheque's collections. For this reason it only includes the data allowing the identification of the samples and their origin, how they were acquired and the dates of their entry and registry. It also includes information about storeroom and cabinet where they are kept so that they can easily be retrieved when they are required. The inventory is completed with a photographic record of each sample.

During the inventory process, the samples are classified in three groups: 1) samples that cannot be lent to third-parties or destroyed, but must be stored whole and indefinitely in the lithotheque; 2) samples that cannot be lent out but which may be destroyed in laboratory preparations and tests; and 3) those that may be lent out.

The first two groups include all the samples forming the core of the lithotheque's reference collection, and which ensure it is exhaustive and representative, and which therefore should be conserved for an indefinite length of time and be available to all the users. Belonging to these groups are the referenced samples at each outcrop, those described in petrological characterisation files, and all kind of preparations and remains taken from them (thin sections, powder preparations, cubes for magnetism measurements, and so on). The third group includes all the other samples that can be lent and exchanged with other groups and users, and are only subject to the general conditions of the service and the regulations governing the use of the collections.

All the samples are washed, weighed, labelled and packaged individually in polypropylene bags to guarantee their conservation, as would be done at any museum, and are stored in standardised plastic containers. The largest and heaviest samples, which are only consulted occasionally, are kept in a storeroom, while the laboratory preparations and a selection of smaller reference materials, which are viewed more frequently, are kept in glass cabinets for easy access by users in the consultation rooms and laboratory.

Information Management

The formation of a lithotheque, like any other archaeological research project, generates a large volume of data in different formats. In a project like this there are, for example, field and laboratory notes for the outcrops recording as well as data from the petrological characterisation of the rocks, photographs, stratigraphic diagrams, maps and a long list of other documents. One of the biggest challenges of the project, on which its success largely depends, is an awareness of the diversity of the information and the complexity of its management.

All the information documenting the collections is digitalised, if it was not already so, and a copy is kept in the computer servers in our institution, which host the project's databases, digital repositories and, shortly, its website. The data which is regularly structured, like the lists of outcrops, the geological units and types of flint, are stored in a relational database system, working with PostgreSQL. This open database management system, in addition to the common utilities in this type of programme, allows geographical objects (extension PostGIS) to be stored and operated with, which is of particular interest in a project producing and distributing information of this kind. Photographs, data records and other kinds

of documents (reports, publications, etc.) which cannot be treated in the same way are stored in digital repositories allowing searches and efficient retrieval.
One particularly important kind of document is the form used to record and describe the flint outcrops. For this, a dynamic digital form is used allowing the information that describes them to be stored in a structured way and export it into an extensible markup language (XML), so that it can be filed in the project's databases.
The form has been especially designed to gather information about a wide range of data which are summarised in a conventional way in the stratigraphic columns used by geologists in their studies and publications (Graham 1998). This procedure is particularly appropriate for the sedimentary nature and emplacement of the rocks being studied. For each rock stratum, its design allows the depth, lithology and identification code of every sample to be recorded in such a way that each outcrop can be referenced with great precision and re-visited if necessary to collect new samples.
The fields in the form are grouped into three blocks. The first includes the basic data identifying the outcrop, in a similar way to other kinds of forms, such as the name, location, access, geological context and outcrop typology. The second group, which is specific to this form, contains the fields in which the local stratigraphy and the siliceous rocks there may be in each layer are described. Lastly, the fields for recording photographs, samples and the form metadata are grouped in the third block.
The geographic information is also treated specially. This type of information may be points, with the coordinates of the outcrops; linear, describing the access route; and polygonal, with the boundaries of the geological units. The first two types are generated directly by the lithotheque technicians and the third is obtained from digital geological maps produced by public institutions.

## 4. First results and future development of the project

The lithotheque collection consists of over 5000 samples of flint (on 31/12/2012) which amounts to a total weight of about 1.5 tonnes, from as many as 60 different sites in Catalonia, Aragon and the southern French region of Languedoc-Roussillon. The spatial databases record an even larger number of sites, as many as 350, where specialised regional literature cites the presence of siliceous rocks, whose full catalogue is estimated to consist of about a hundred different types.
In 2012 the collections and equipment were installed in the new rooms especially prepared for the purpose at CSIC-IMF (Barcelona), including storeroom, consultation room and laboratories. In the same year, it began to offer its services to researchers linked to the centre and external users who one way or another were aware of the collection contents and other resources in the lithotheque (equipment, maps, bibliography, etc.), although due to the lack of specific funding, its dissemination and the general provision of its services have been delayed until 2014.
Work will continue in order to collect the rock types that are still needed to complete the lithotheque and their characterisation, while the project itself, which is far from complete, will continue developing and re-defining itself in several directions. Two of these can be highlighted.
First, and most urgently, the infrastructures and technological resources supporting the project need to be fully implemented. These include databases, map servers and digital document repositories. Although this is not particularly complicated, it requires study and careful decision-taking. On it depend many of the possibilities of the future development of some key project objectives, such as the efficient offer of a full range of services to remote users via the web and guarantees for full interoperability of its systems, structures and data.
The second, and most ambitious future development, consists of fostering and participating in the creation of a national and international network of lithotheques in order to share data, exchange services and lend them jointly to users in the same point of access. For this, we have reached preliminary agreements by drafting specific collaboration documents with the Universidad del País Vasco/Euskal Herriko Unibertsitatea (UPV/EHU, Vitoria-Gasteiz) and the Centre Européen de Recherches Préhistoriques (CERP, Tautavel). These lithotheques were selected because their collections are complementary to the LitoCAT collection, enabling the integration of geological information in a large circum-Pyrenean area, and the pooling of data about the availability of siliceous rocks in this geographical region.
To conclude, it should be stressed that this project has laid the foundations for a unique research platform. Apart from the intrinsic interest of its research, its potential for the transfer of knowledge and technology to other researchers, teams and institutions is enormous. However, the project's continuity and attainment of its scientific goals cannot be achieved

without the funding needed, above all to recruit full-time technical staff, for the feasibility of the project and its long-term planning towards the scientific success that we would all like.

### Acknowledgments

The Lithotheque Project has been funded by the recruitment programmes JAE-TEC of the Spanish National Research Council and PTA of the Ministry of Science and Innovation, and the financial support of the Culture Department in the Generalitat de Catalunya, whom we thank for their support and funding.

### References

Biró, K.T. and Dobosi, V., 1991. *Litotheca – Comparative Raw Material Collection the Hungarian National Museum*, Budapest: Hungarian National Museum.

Biró, K.T., Dobosi, V. and Chlénder, Z.S., 2000. *Litotheca – Comparative Raw Material Collection the Hungarian National Museum (Vol. II)*, Budapest: Hungarian National Museum.

Biró, K.T., 2011. Comparative raw material collections in support of petroarchaeological studies: an overview", In A. K.T. Biró and. Markó (eds.), *Papers in honour of Viola T. Dobosi*, Budapest: Hungarian National Museum, 225-244.

Borghini, G., 1989. *Marmi antichi, Materiali della cultura artistica*. Roma: De Luca Editori d'Arte.

Graham, J., 1988. Collection and analysis in field data. In M.E. Tucker (ed.), *Techniques in Sedimentology.* Oxford: Blackwell Scientific Publications, 5-62.

Herz, N. and Waelkens, M. (eds.), 1988. *Classical marble: geochemistry, technology, trade*. Berlin: Springer.

Prikryl, R. and Török, Á. (eds.), 2010. *Natural Stone Resources for Historical Monuments*. London: Geological Society (Special Publications, 333).

Suárez, A. and Tsutsui, N.D., 2004. The Value of Museum Collection for Research and Society. *BioScience*, 54/1, 66-74.

Terradas, X., Álvarez, A., Bartrolí, R., Borrell, F., Clop, X., Fullola, J.M., García-Antón, M.D., Gibaja, J.F., Gómez, B., Grégoire, S., Mangado, X., Maroto, J., Martluff, M., Molist, M., Ortega, D., Palomo, A., Sala, R., Soler, N. and Vila, A., 2006. Red temática «Estudio de la disponibilidad de rocas silíceas para la producción de instrumental lítico en la Prehistoria». In G. Martínez, A. Morgado and J.A. Afonso (eds.), *Sociedades prehistóricas, recursos abióticos y territorio*. Loja: Fundación Ibn al-Jatib de Estudios de Cooperación Cultural, 63-72.

Terradas, X., Ortega, D. and Boix, J., 2012. El projecte LitoCAT. Creació d'una litoteca de referència de roques silícies de Catalunya. *Tribuna d'Arqueologia*, 2010-2011, 131-150.

Turq, A., 2005. Réflexions méthodologiques sur les études de matières premières lithiques. 1.- Des lithothèques au matériel archéologique. *Paléo*, 17, 111-32.

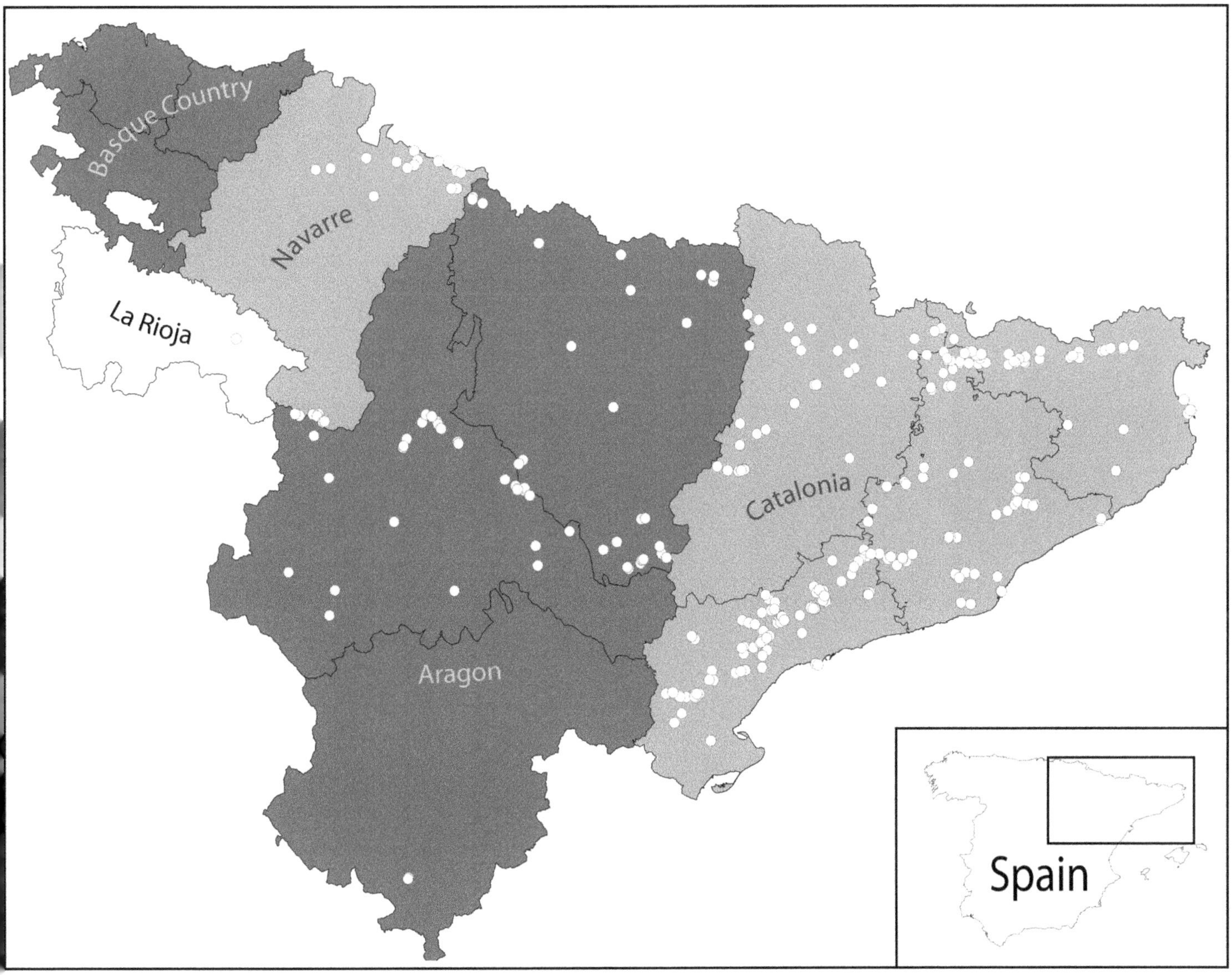


Fig. 1 : Map of the north-eastern Iberian Peninsula marking the flint outcrops recorded in the lithotheque's databases (updated in July 2012).

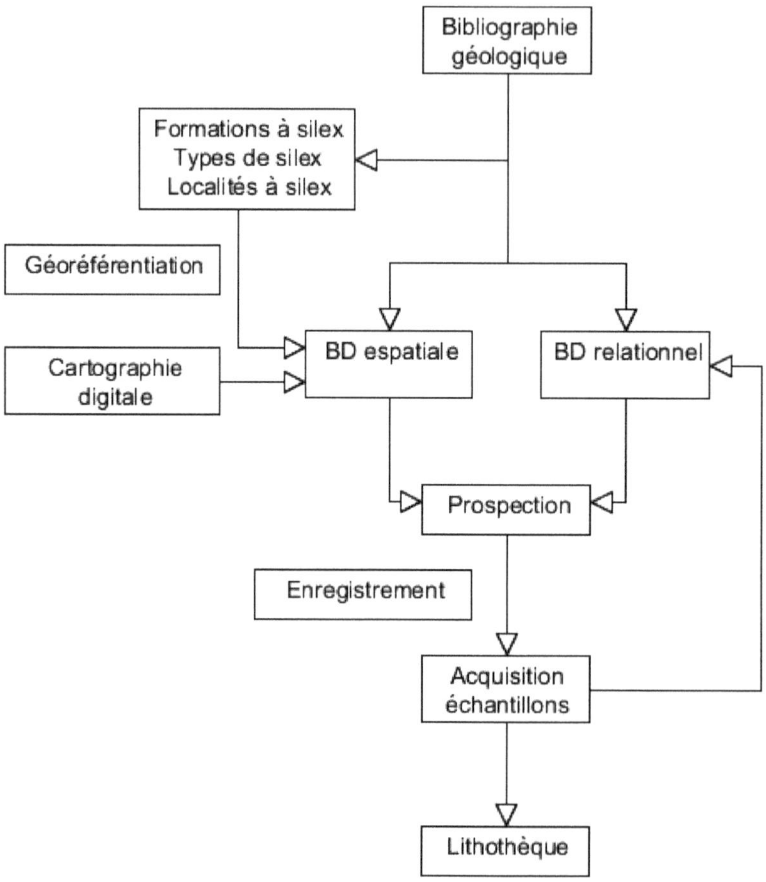


Fig. 2 : Lithotheque work flowchart.

# A siliceous rocks database for the northeastern Paris basin : the Soissons lithotheque

Claira Lietar, Olivia Dupart, Bruno Robert, François Giligny, Françoise Bostyn, Laurence Manolakakis

**Abstract**

The Soissons reference collection has been created in the late 1980 s, in a region, in which the Early/Middle Neolithic period has been extensively investigated. The current French-German MK-project has provided the opportunity to update the existing collection and to incorporate the information into a GIS. The database contains 440 rock samples, mainly from the Aisne and Oise drainage systems in the north-eastern part of the Paris basin. Non-systematic sampling has now extended to the entire Paris basin and to neighbouring regions, including data from the Petite Seine, the Marne and the A5 motorway sites, as well as flint from the Jura Mountains, the Lorraine region, Poland, Belgium, Holland, the Rhineland and Hungary. In parallel, a similar database for clay is under construction.
The first results concern the relationships between lithic technology and flint/stone raw materials in the Neolithic. The lithothèque will possibly provide a reference collection for the identification of raw materials in the Paris basin.

**Keywords :** reference collection; Lithotheque; flint; procurement strategy; Paris Basin; Neolithic.

## 1. Historiography and aims

The project of a reference collection or lithothèque has been developed within an ATP (Action Thématique Programmée) between 1986 and 1989 by scientists of the Archaeological Research Centre (Centre de Recherches Archéologiques) of Soissons, as well as of the associated laboratories (« Département des Sciences Naturelles » at the University of Reims and the « Laboratoire d'étude de l'environnement » at the Institute of Prehistory of the Leiden University, Netherlands). Fieldwork has been carried out by a network of surveyors and archaeologists, the principal actors being Michel Plateaux, Jean-Claude Blanchet, Claudine Pommepuy, and Gilles Naze (Blanchet *et al.* 1989).

This project was part of a research theme on the procurement strategies of lithic raw materials and supply territories of the neolithic and protohistoric sites in the Aisne valley. It primarily aimed at the systematic sampling of local geological formations within the catchment area of the Aisne valley (departments of Aisne and Oise) and the seeking for the potential raw material sources based on already petrographically determined archaeological objects. The sampling method is based on the systematic analysis of 1:50.000 scale geological maps and on the description of each formation in an analytical sheet (description of the formations and of the material types by geological stage). In Picardy, the entire Aisne and Oise departments as well as one part of Champaign (Ardennes, Marne), i.e. thirty geological maps, have been analysed, thus allowing a rapid evaluation of the quality and the variability of the existing materials and the identification of the most interesting formations.

Sampling has then been made in mines or on outcrops for each of the described formations with at least one sample, except for limestone, which is overrepresented, and for which only the larger petrographic type has been determined (coarse-grained limestone, cerithoid limestone, shelly limestone, miliolids limestone, nummulitic limestone). The potential of each sector has been evaluated in more detail. Finally, sampling has been performed during systematic surveys in sectors of particular interest for local surveyors : the mines and knapping workshops in the region of Nointel (P. Romenteau) ; the formations in the surroundings of Cuiry-lès-Chaudardes (Y. Naze, M. Plateaux), of Boury-en-Vexin (D. Verret, R. Martinez), the Ludian formations in the Tardenois sector (M. Plateaux, C. Pommepuy), and in the area of the Aisne river (M. Plateaux) as well as in gravel pits in the surroundings of the site of Amigny-Rouy (Y. Naze).

The assemblage has finally been enlarged in a non-systematic manner to the entire Paris basin by incorporating data stemming from La Petite Seine (A. Augereau), the Marne (F. Bostyn), the Oise (F. Prodéo) and from the sites discovered during the survey of highway A5 (P.A. de Labriffe). Samples stemming from the adjacent regions of the Paris basin have also been incorporated into the database, still in a non-systematic manner (Jura, Lorraine, Belgium, Netherlands, Rhineland), or from more distant countries (Poland and Hungary). Since, an additional research project has been developed on Secondary flint, «*Géoarchéologie du silex du Nord-Ouest de la France*» (Allard *et al.*, 2005). The materials sampled within this research program have been shared by those of the Soissons reference collection and they are kept at the INRAP in Amiens.

The petrographical determination based on thin sections of all the materials has been performed by two geologists at the University of Reims (H. Guérin and J.-C. Foucher). It has notably permitted the determination of the sandstones and the distinction between Secondary flint (Upper Cretaceous) and Tertiary flint through micropalaeontological analysis of the enclosed fossils. The accurate attribution to a distinct geological stage that allows dating is however rare and only possible for very characteristic samples (presence of diagnostic fossils).

In a more general manner, the sampling, for the greater part, is representative of the northern half of France; the best represented sector being the north-eastern quarter of the country (Fig. 1 & 2). The assemblage extends eastwards, including samples stemming not only from neighbouring countries but also from more distant areas (mainly Belgium and Germany, the Netherlands and Poland). Only few samples stem from Southern and Western France.

The reference collection has permitted to carry out an intensive assessment of the stone resources in the Aisne-Vesle sector and in the central Oise valley principally (Fig. 2) : flint supply stemming from primary formations of the Mesozoic and the Tertiary, is composed, in the eastern part of the investigated area, of black Turonian flint of Rethel and greenish Cretaceous flint in the area of the Ile-de-France cuesta, of Senonian flint from Epernay in the southern part and of flint stemming from the Santonian and Campanian chalk formations west of the Oise valley. The flint resources stemming from primary formations of the Tertiary are located in the levels of the Upper Lutetian on the edges of the plateaus surrounding the Aisne valley. The formations of the Upper Bartonian (Ludian) are situated on the Tardenois plateaus. The residual formations on the terraces of the Aisne valley yield nodular quartz sandstone (Auversian), pebbles of Turonian flint as well as Tertiary silifications. In addition to the sampling of these formations, the analysis of the geological maps and surveys has enabled the discovery of new sources in the local range as those of « quartzite sandstone » (sandstone constituted by over 80% of quartz grains or with mainly siliceous cementation). Its existence had been assumed by the archaeologists but remained unknown on the regional level. Their texture makes them convenient for knapping.

## 2. The database

The relational database has been created with Microsoft Access. It is subdivided into four sections. The first section concerns the information allowing to localize a piece in the reference collection and in the database, i.e. the number of the drawer corresponding to the original storage in the cupboards, the box number which corresponds to the new conditioning of the boxes made in 2012, the number of the card corresponding to the unique identifier of each registration in the database as well as the date of registration. A second section concerns the field data, i.e. the commune and perhaps the locality of the discovery, the name of the surveyor and the Lambert coordinates 1 and/or 2 as well as the number of the geological map and the numbering of the sites and the samples from 1 to n following the registration order of the pieces. The third section concerns the geological information of the sample, i.e. its identification as a raw material (limestone, sandstone, obsidian, flint, silification...), the geological stage and sometimes the precise code of this stage. The modalities of access to the materials (outcrop, mine, extraction pit, test trench or archaeological site) and their quantity on the site (abundant, medium or rare) are also described. Finally, two systems of measurement are proposed, a range and an average. The last section concerns the physical description of the raw material: the colour, the grain, the physical state of the piece, as well as the aspect of the cortex and finally the presence or absence of microscopic or macroscopic fossils.

Within the French-German ANR-DFG project « Emergence of social complexity: enclosures, resources and territoriality in the Neolithic. French-German research on the Michelsberg culture » the reference collection of Soissons has been updated. The database of the reference collection currently contains two hundred and sixty-eight samples, distributed over forty-seven geological maps. Three maps only show more than fifteen samples: the map of Soissons (17 pieces), the one of Clermont (50 pieces) and the one of Saint-Just-en-Chaussée (44 pieces). The distributions by stage are utterly representative and consistent with the geological maps. Considering all materials, the database now contains four hundred and forty-four cards of which two hundred and eighty-five are described. One hundred and ninety-four samples, mainly flint have in addition been photographed, and this work will soon be completed for sandstone and limestone.

## 3. The raw materials

The major part of the samples registered in the database stem from the Cretaceous and the beginning of the Tertiary (Eocene and Palaeocene; Fig. 3). The raw materials of the Jurassic and the Triassic periods are only represented by about fifteen samples. The materials are presented here by the larger geological periods and the physical criteria of the different geological stages are identified and described. Only well characterized samples are used for this purpose. Deposits in secondary position or materials found in silt layers have been excluded as well as those which have not been well differentiated between two geological stages. Finally, all the pieces found in alluvial contexts or in silt contexts have not been taken into account, in order to base on most reliable data.

Three descriptive criteria have been chosen for the description of the materials: the morphology, the original colour and the thickness of the cortex. These three criteria have been selected in that they allow to have a global picture of the raw material and in that their description is standardized. We are aiming at proposing recurrent general criteria and do not attempt a typology of raw materials, given that the variation within a same geological facies is too important. We also have to take into consideration biases in sampling (number of pieces by deposit and representativeness of the occurring varieties). Finally, it is quite clear that the proposed macroscopic observations cannot substitute to petrographic analysis.

The Tertiary flint materials

The assemblage contains sixty pieces identified by primary geological stages. All the geological stages are represented, from the Ypresian to the Ludian of which the registrations are grouped together under the term « Upper Bartonian ». However, the upper stages are the best represented in the reference collection: twelve samples for the Lower-Middle Bartonian, fourteen for the Upper Bartonian (Ludian) and nineteen for the entire stage of the Lutetian (fig. 4). On the other hand, the earlier stages are sampled with at least ten elements: three for the Sparnacian, nine for the Thanetian and only one for the Ypresian. Finally, two pieces assigned to the « Sparnacian/Thanetian » have been considered for the distribution map but they are set aside for the description. The raw materials exclusively stem from the north-eastern part of the Paris Basin. The best sampled area is located north and south of the Aisne River, between Soissons and Reims.

The distribution of the morphologies under the form of tabular or nodular flint allow to highlight several facts (Fig. 5 et 6): it can be stated that the raw materials which occur mainly under the form of tabular flint stem from the upper stages of the Eocene, i.e. the Ludian and the Bartonian: within the reference collection, the uppermost stage is exclusively represented by small slabs, as well as by a single bloc and a slab. The lower stages of the Bartonian contain a majority of tabular flint but also some nodules. Within the Lutetian stage, an intermediate stage of the Eocene, nodular and tabular flint occur in equal parts; moreover, it can be noted that the proportion of samples containing both tabular and nodular flint almost equals the preceding categories. Lastly, the Thanetian is mostly represented by nodules, occasionally by a pebble and not a single tabular flint piece occurs in the assemblage.
We have only three samples for the Sparnacian and one for the Ypresian. These elements do not allow to draw valid conclusions. It can merely be noted that the distribution of the Sparnacian overlaps with the elements of the Thanetian.

The thicknesses of the cortex of the samples equally mirror general trends (Fig. 7). With regard to the upper stages, a distinction between the two most recent stages can be noted : thus, on the samples of the Ludian, in most cases thick cortex between 5 and 10 mm can be observed, whilst it is thinner (1-2 mm), and even ultra-thin for the lower and middle stages of the Bartonian. Once again, the stages of the Sparnacian and the Ypresian are too insufficiently represented to draw any conclusions.

Only the middle and upper stages of the Eocene show a large variety of colours. More than 50% of the Ludian samples show red colours ranging from « pink light brown » to « brown dark red » that characterize 35 % of the registered pieces. The other colours are brown or grey with rather light nuances (Fig. 8).

Within the Lutetian stage, 20% of the samples show red colours but the greater part is comprised within brownish shades of colour (brown, light brown, dark brown, dark grey-brown totalizing 60 %). The shades of colour close to light and dark grey-brown complete the range (Fig. 8). In the Bartonian stage the major shades of colour range from brown to dark grey-brown (44%). 24 % of the samples show significantly lighter tints, the red colours represent only 16% of the assemblage (Fig. 8). For the lower stages, the variability of colours seems more restricted. Only the Sparnacian stage still provides enough elements: the tints vary from light grey to dark grey. The samples of the Ypresian and the Sparnacian do not distinguish from these tints (Fig. 8).

The Cretaceous flint materials

The assemblage contains ninety-two localized elements. The sampled area is centred on the north-eastern part of the Paris Basin concentrating in the region covered by the maps of Saint-Just-en-Chaussée, Beauvais and Clermont.
Almost all the geological stages of the Cretaceous are represented with thirty-one samples for the Campanian, nineteen for the Santonian, twenty-nine for the Coniacian and six for the Turonian. For seven samples it was impossible to determine the geological stage: six are registered as « Santonian/Coniacian » and one as « Turonian/Coniacian » (fig. 9). The Maastrichtian flint, the most recent stage of the Cretaceous, is represented by two samples only, one stemming from the commune of Eben-Emaël in Belgium, the other from the mining area of Rijckholt in the Netherlands.
The samples registered under the generic term « Cretaceous » have not been taken into account.

By contrast to the morphologies of the Tertiary age, those associated with the siliceous materials of the Cretaceous vary little: almost the totality of the samples occurs as nodules. The presence of tabular flint is anecdotic for the stages of the Campanian, the Santonian and the Coniacian (Fig. 10 and 11).

Equally, the greater part of the pieces presents an ultra-thin or thin (1-2 mm) cortex, for all of the materials. Only the samples of the Coniacian distinguish through thicknesses in average more important compared to the others and through a larger proportion of medium thick cortex (2-5 mm). The absence of very thick cortex (> 10 mm) can also be noted for the present assemblage (Fig. 12).

The colours of the Cretaceous flints vary from beige to very dark grey. The samples are mostly comprised within « brown-grey » or « grey-brown » tints. The very dark grey flint is present in all geological stages. The lighter shades of grey or brown are little represented (Fig. 13).

Thus, there are no clear distinctions between the Cretaceous stages based on the three criteria we have examined, by contrast to the observations made on Tertiary flint and silifications that show clear differentiation, even an evolution concerning the aspect of the raw materials according to the geological stages. From a physical point of view, it thus seems difficult to assume the geological attribution of the Cretaceous flints.

The Jurassic and Triassic flint materials

Out of a total of sixteen flint samples from both the Jurassic and Triassic periods, fourteen are localized. The sampling of these materials thus remains anecdotic for the reference collection of Soissons. In Normandy, the raw materials belong to the Middle Jurassic period; the two samples of the Bathonian stem from the mines of Ri and of Bretteville-le-Rabet, those of the Bajocian from Moutiers-en-Cinglais. The remaining samples of shelly limestone stemming from the Upper Jurassic (Oxfordian) or the Middle Jurassic are localized in Eastern France in an area spanning from the Ardennes in the north to the city of Epinal in the south.

Despite the apparent weakness of the sampling it is nonetheless possible to highlight distinct characteristics of the Triassic and Jurassic raw materials (Fig. 12). Thus the shelly limestone flints present within the reference collection exclusively occur under the form of tabular flint (5 specimen), whilst the Jurassic raw materials are found rather in the form of blocs or nodules, possibly pebbles. The thickness of the cortex is a criterion which allows to highlight that the Jurassic as well as the Triassic flint materials present mostly thin to ultra-thin cortex. On the other hand, the two samples of the Bathonian (Middle Jurassic) have thick to ultra-thick cortex (5-10 mm and > 10 mm). Finally, the colours of the Jurassic and Triassic flints span from light to very dark grey; one sample of the Middle Jurassic (Bajocian) has a « very dark grey brown » colour. In a more general manner, the colours of the Jurassic pieces seemingly exhibit lighter nuances than those of the Triassic period.

## 4. Case Studies

The Soissons reference collection has been used to perform several studies on raw material procurement strategies.

First and foremost, the research conducted by one of the initiators of the reference collection, M. Plateaux, into the Neolithic lithic industries of the Aisne valley and the Paris basin has to be mentioned (1987, 1989). More recent studies pursuing this work are based on the reference collection, for example those by P. Allard (2005). This latter has demonstrated an important variability of the types of procurement in the Early Neolithic Linear Pottery sites (5200-4900 BC), with sites rather close the ones to the others but showing quite different proportions of both local and regional materials. In these sites, the procurement is local and regional: the used materials are local Tertiary flint, Turonian flint, and Senonian flint from the more than 50 km distant Marne. The raw materials are mostly stemming from Mesozoic formation (chalk flint). Although the accessibility of the site is the same a priori, the strong variability of the materials from one site to the other apparently is due to cultural choices (Allard 2005).

With regard to the sandstones, another founder of the reference collection, C. Pommepuy, has carried out a synthesis on the grinding stones in the Bronze Age and the Iron Age in the Aisne valley. She has demonstrated choices in the use of distinct materials, in a region displaying a large number of diversified resources: sandstone and more particularly Belleu sandstone in the Final Hallstatt and Early La Tène period, limestone in the recent and final La Tène period (Pommepuy 1999).
In their studies, C. Hamon and G. Fronteau have also proposed a synthesis on sandstone materials outcropping in the Aisne valley from east to west, between the Champagne plain and the Soissons region (Hamon and Fronteau 2010). They have highlighted a large variety of accessible materials of which four at least are used during the Linear Pottery Culture.

At the end of the 5th millennium, in the area of the Michelsberg culture, the sites are not systematically established next to the flint resources and procurement focuses on high quality materials. For example, the sites of the Aisne valley indicate a major procurement of Bartonian flint from Romigny-Lhéry (at a distance of about 20 to 50 km) and an important supply with Senonian flints, probably stemming from the region of Epernay in Champagne. Here, debitage, although primarily on flakes, is oriented towards a true laminar production of two types: a « domestic/household » production of blades made from both Senonian and Bartonian flint and a production of large blades made on Bartonian flint, but also on Spiennes and Rijckholt flint, possibly processed in mine sites of the Michelsberg area (Giligny & Manolakakis 2011).

## 5. Perspectives

After the updating of the data base and the description of the samples conducted within the French-German « MK-project », the reference collection currently is a useful tool for the study of lithic materials recorded from archaeological sites of the north-eastern part of the Paris basin. The development of new laboratory analyses may be envisaged as new petrographic analyses or those made with other methods.

The reference collection is considered as a referential in progress that can be enlarged by new surveys. Distinct areas would be worth investigated in a more systematic manner, for example the plateaus of the Tardenois and the central Oise valley, and in a broader context the Seine valley.

Materials other than flint, for example clays, limestone and sandstone will also be subject to a reassessment of the data and their descriptions completed in the database in order to complete the referential of lithic resources of the study area. An extension of the reference collection is planned through the incorporation of clay samples of the Aisne valley collected by one of us in the last years (B. Robert). Thin sections of backed samples will also be made soon and incorporated into the reference collection and used for pottery studies.

In the short term, the principal information of the reference collection will be placed on-line. About hundred raw material sources will be described including for each the localization of the site (administrative and geographic), the geological attribution and the macroscopic description of the samples.

The Soissons Lithotheque is a useful referential for studies on numerous materials (limestone, sandstone, flint) which in the future may be part of a network of regional reference collections as those of Amiens, Normandy and Belgium constituting a referential for the entire Paris basin and its margins.

**Acknowledgments**
This project was funded by the ANR-DFG MK-Projekt (2010-2013).
Translation: K. Mazurié de Keroualin

## References

Allard, P. 2005. *The stone working of Linear Pottery populations in north-eastern France and Belgium,* Internationale Archäologie, Rahden/Westf, 290 p.

Allard, P., Bostyn, F., Fabre, J., 2005. Origine et circulation du silex durant le Néolithique en Picardie. Des premières approches ponctuelles à une systématique régionale, *Revue archéologique de Picardie*, Numéro spécial 22, 49-74.

Allard, P., Bostyn, F., Martial E. 2010. Les matières premières siliceuses exploitées au Néolithique moyen et final dans le Nord et la Picardie (France), In C. Billard (ed.) *Premiers néolithiques : cultures, réseaux, échanges des premières sociétés néolithiques à leur expansion dans l'ouest de la France*, actes du colloque interrégional sur le Néolithique au Havre, novembre 2007, Presses Universitaires de Rennes, 355-383.

Allard P., C. Hamon (eds.), S. Bonnardin, N. Cayol, M. Chartier, A. Coudart, J. Dubouloz, G. Fronteau, L. Gomart, L. Hachem, M. Ilett, K. Meunier, C. Monchablon, C. Thevenet, 2010. *Économie et société des populations rubanées de la vallée de l'Aisne*, Rapport final d'activité du PCR 2007-2010, Service Régional de l'Archéologie – DRAC Nord-Picardie Amiens.

Blanchet, J.-C., Plateaux, M., Pommepuy, C. (eds.), 1989. *Matières premières et sociétés préhistoriques dans le Nord de la France*. Rapport final d'activités, Action Thématique Programmée « Archéologie métropolitaine », Direction des Antiquités de Picardie, Soissons, 76 p.

Giligny F., Manolakakis L., 2011. Territories and lithic resources in the Paris basin during the middle Neolithic (4200-3600 BC), in M. Capote , S. Consegrua, P. Diaz-del-Rio, X. Terradas (ed.) *Proceedings of the 2nd International Conference of the UISPP Commission on Flint Mining in Pre- and Protohistoric Times*, Madrid 14-17 oct. 2009 , 45-50.

Plateaux, M., 1987. L'industrie lithique des premiers agriculteurs dans le nord de la Franc. In : Stone Industry in the Early Neolithic of Europe, Archaeologia interregionalis, Warsaw, 225-245.

Plateaux, M., 1990. Approche régionale et différentes échelles d'observation pour l'étude du Néolithique et du Chalcolithique du Nord de la France ; l'exemple de la vallée de l'Aisne, In coll., *Archéologie et espaces,* Xe Rencontres internationales d'Archéologie et d'Histoire, Antibes, octobre 1989, APDCA, Juan-les-Pins, 157-182.

Plateaux M, 1989. Contribution à l'élaboration d'une problématique des matières premières pour le Néolithique du Bassin parisien. In J.-C. Blanchet, A. Bulard, C. Constantin, D. Mordant, J. Tarrête (eds.), *Le néolithique au quotidien.* Actes du XVIe Colloque Interrégional sur le Néolithique, Ed. de la Maison des Sciences de l'Homme, Paris, 1993, p. 100 à 104. (Documents d'Archéologie de Française).

Pommepuy C., 1999. Le matériel de mouture de la vallée de l'Aisne de l'âge du bronze à La Tène finale : formes et matériaux, *Revue Archéologique de Picardie*, 1999, 3, 115-141.

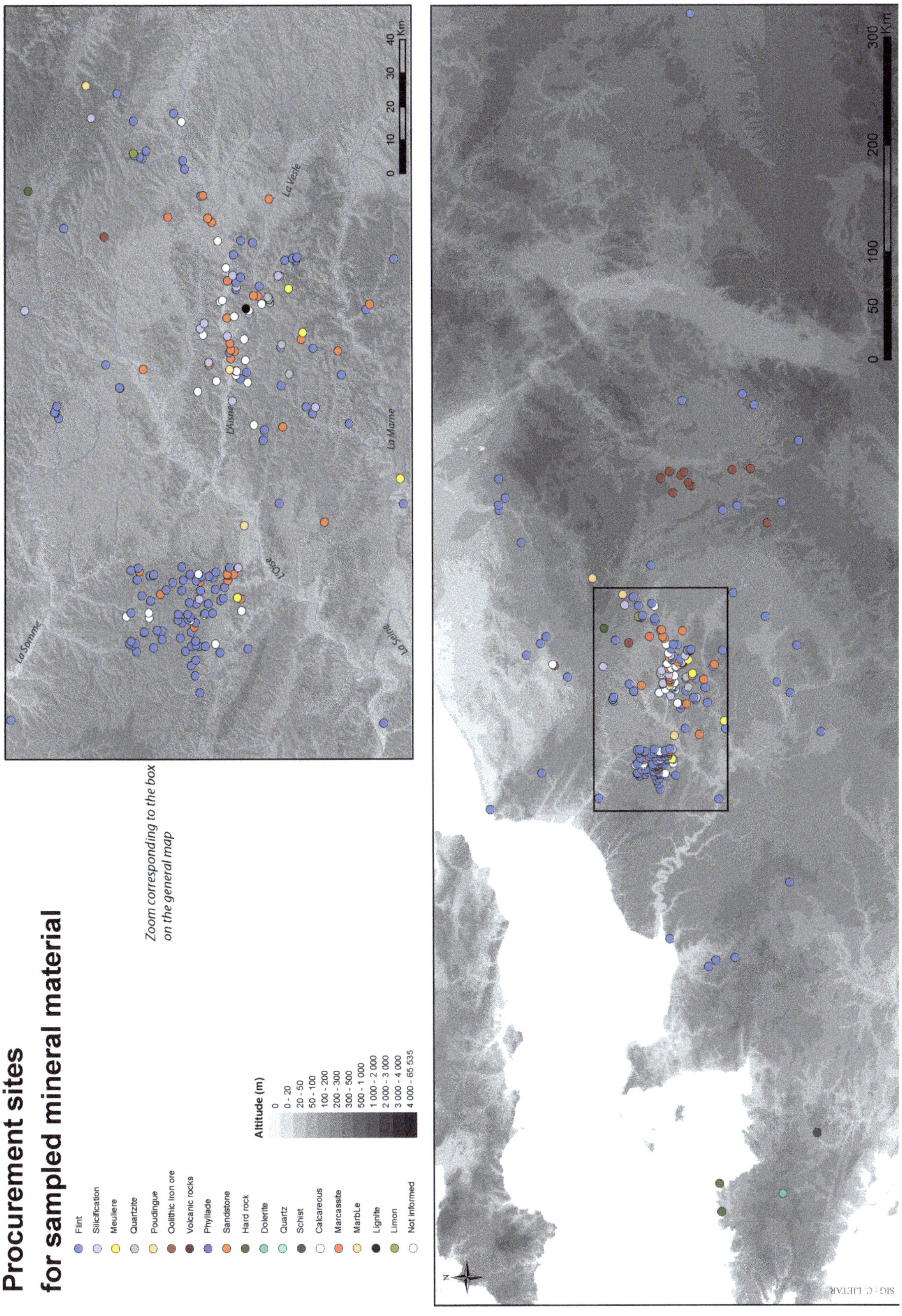


Fig. 1 : Distribution of the sources documented in the Soissons reference collection by material in Northern France and the adjacent regions.

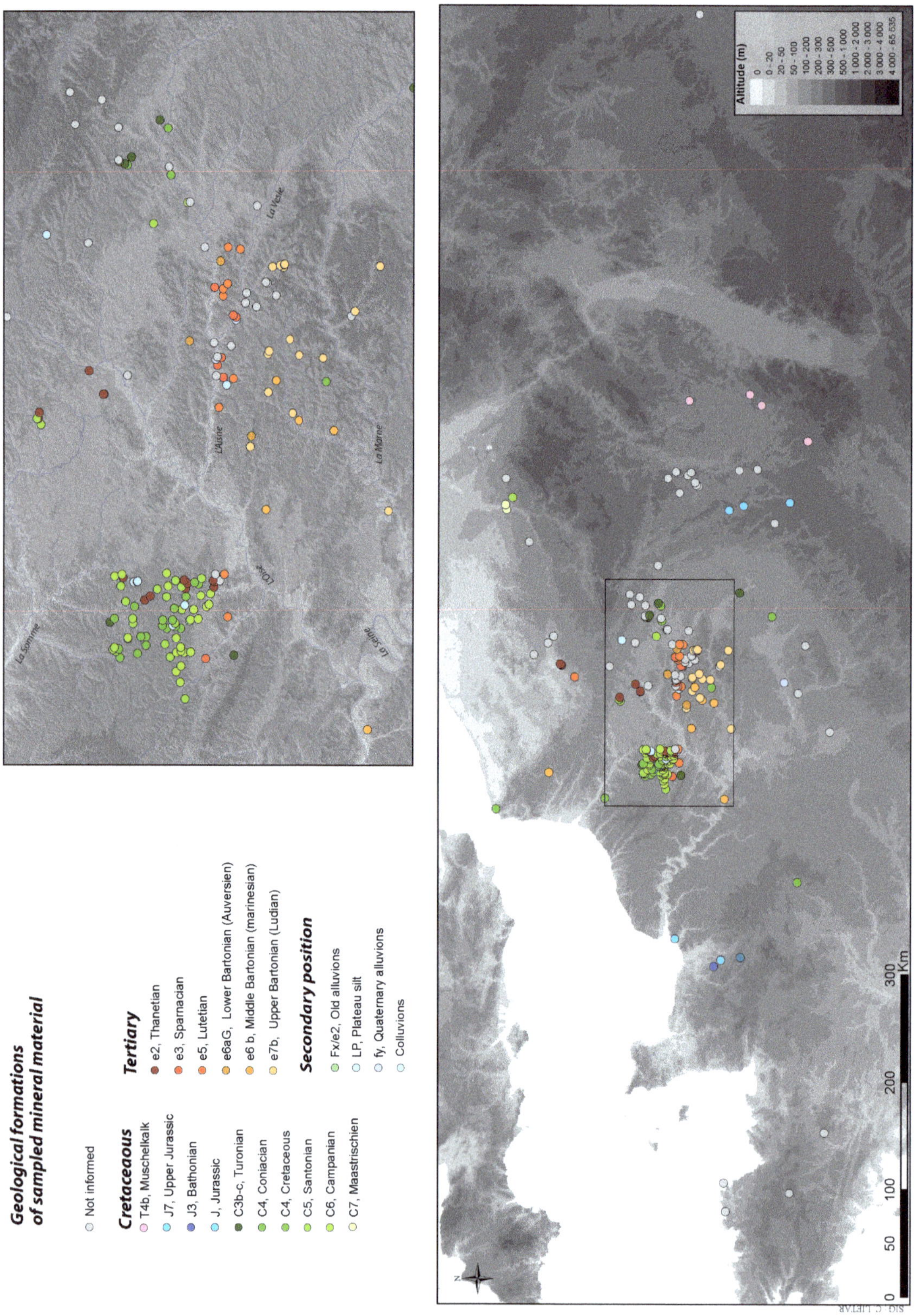


Fig. 2 : Distribution of the sources documented in the Soissons reference collection by geological stage in the Aisne and Oise valleys

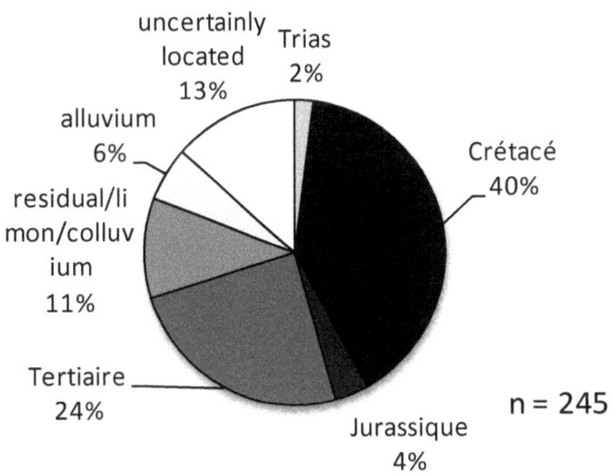


Fig. 3 : Distribution of the flint raw materials and silicifications by geological age.

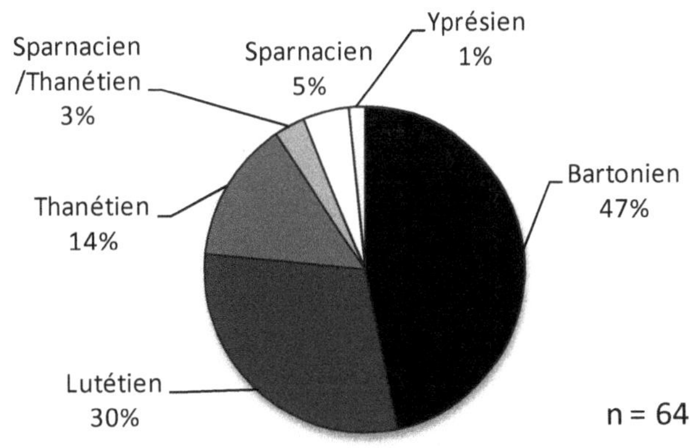


Fig. 4 : Distribution of the flint raw materials and silicifications of the Tertiary by geological stage.

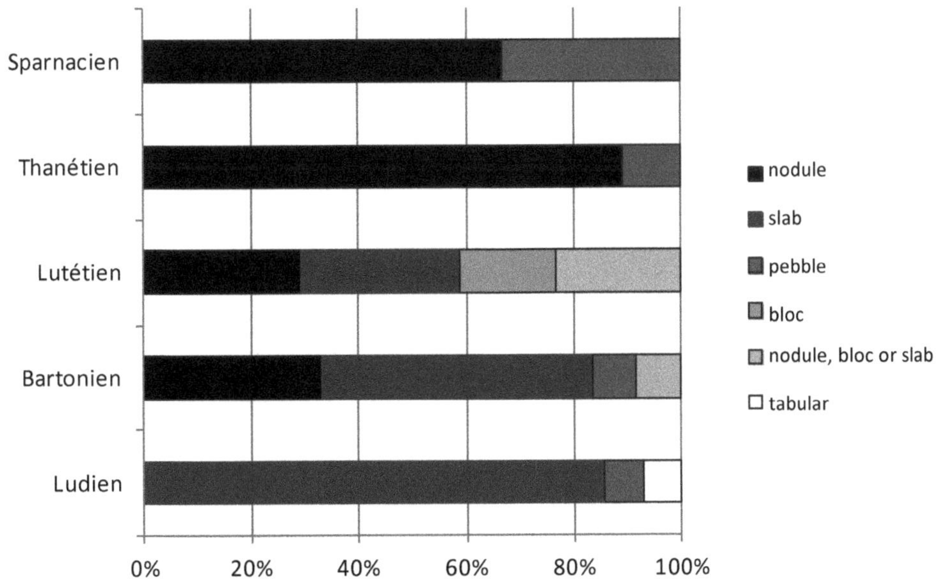


| | nodule | slab | pebble | bloc | nodule/ bloc/ slab | tabular | Total |
|---|---|---|---|---|---|---|---|
| Ludian | | 12 | 1 | | | 1 | 14 |
| Bartonian | 4 | 6 | 1 | | 1 | | 12 |
| Lutetian | 5 | 5 | | 3 | 4 | | 17 |
| Thanetian | 8 | | 1 | | | | 9 |
| Sparnacian | 2 | | 1 | | | | 3 |
| Ypresian | | 1 | | | | | 1 |
| | 19 | 24 | 4 | 3 | 5 | 1 | 56 |

Fig. 5 : Morphology of the samples of the Tertiary by geological stage.

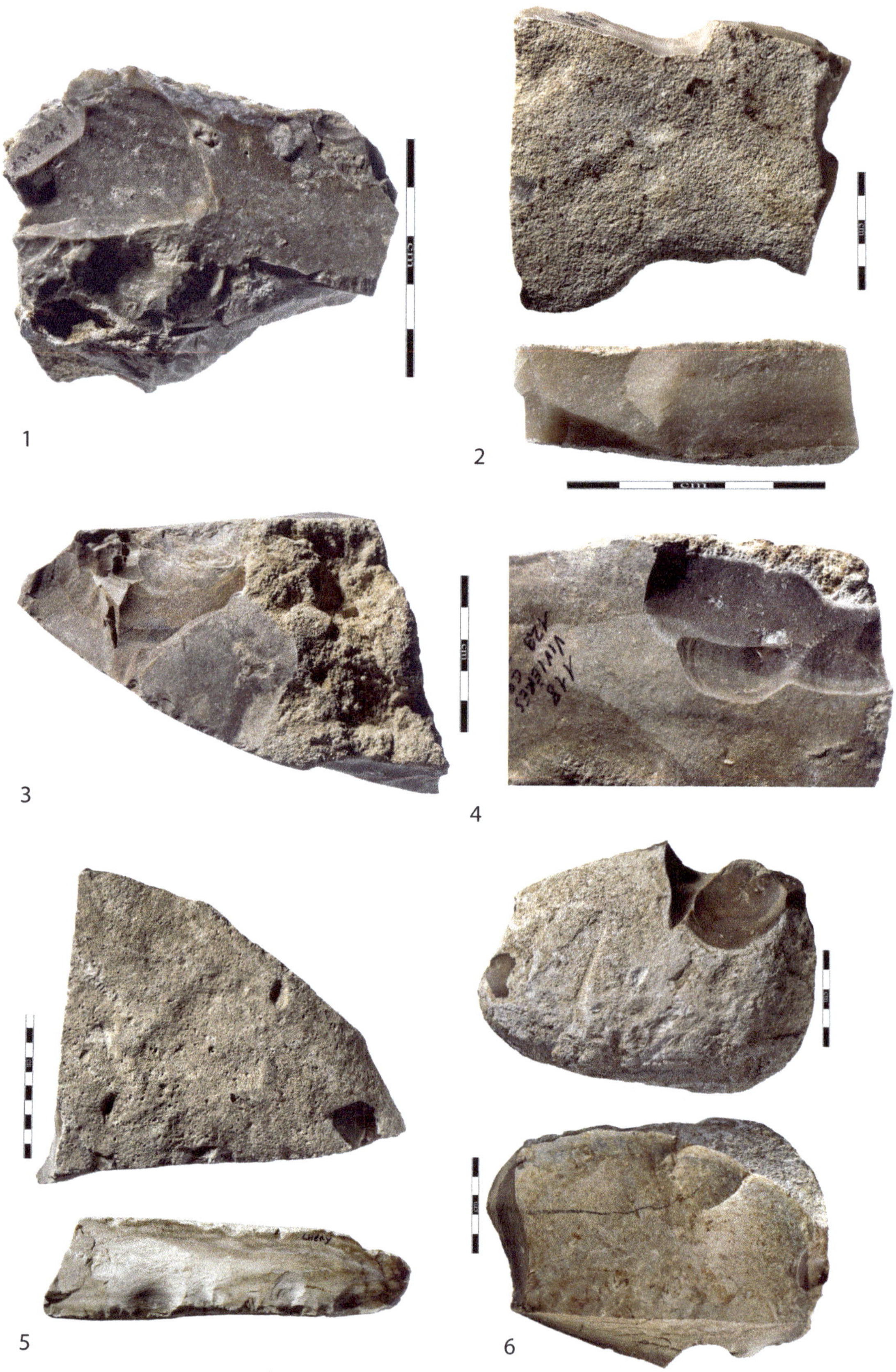

Fig. 6 : Silicifications of the Tertiary. 1. Glennes, Lutétien e5, 2. Beugneux, Ludien e7a quartzite, 3-4. Vivières, Bartonien e6, 5. Lhéry, Bartonien supérieur e7b , 6. Fl–ins-sur-Seine, Bartonien moyen/supérieur e6b-e7. (photos S. Oboukhoff, CNRS).

| | pellicular | 1-2 mm | 2-5 mm | 5-10 mm | > 10 mm | irregular | Total |
|---|---|---|---|---|---|---|---|
| Ludian | 2 | 2 | 1 | 6 | 2 | 2 | 15 |
| Bartonian | 3 | 7 | 1 | 1 | 1 | | 13 |
| Lutétien | 3 | 3 | 1 | 4 | 2 | 7 | 20 |
| Thanetian | 7 | | | | | 1 | 8 |
| Sparnacian | 2 | | 1 | | | | 3 |
| Ypresian | | | | | | 1 | 1 |
| | 17 | 12 | 4 | 11 | 5 | 11 | 60 |

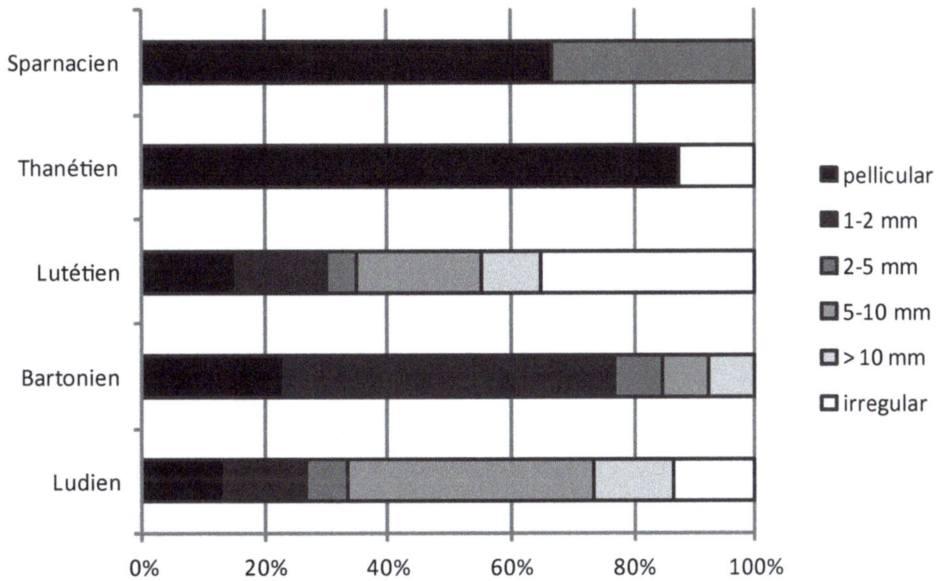


Fig. 7 : Cortex thickness of the Tertiary samples.

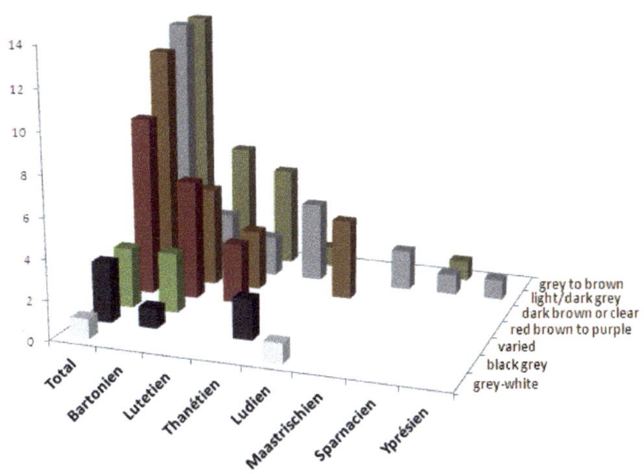


| | grey to brown | light/dark grey | dark brown or clear | red brown to purple | black grey | grey-white | varied | total |
|---|---|---|---|---|---|---|---|---|
| Yprésien | | 1 | | | | | | 1 |
| Sparnacien | 1 | 1 | | | | | | 2 |
| Maastrischien | | 2 | | | | | | 2 |
| Ludien | | | 4 | | | 1 | | 5 |
| Thanétien | 1 | 4 | | | 2 | | | 7 |
| Lutetien | 5 | 2 | 3 | 3 | | | | 13 |
| Bartonien | 6 | 3 | 5 | 6 | 1 | | 3 | 24 |
| Total | 13 | 13 | 12 | 9 | 3 | 1 | 3 | 54 |

Fig. 8 : Distribution of the colours of the silifications of the Tertiary.

| | |
|---|---|
| Campanian | 31 |
| Santonian | 19 |
| Santonian/ Coniacian | 6 |
| Coniacian | 29 |
| Turonian/ Conacian | 1 |
| Turonian | 6 |
| **total** | 92 |
| Maastrischian | 2 |

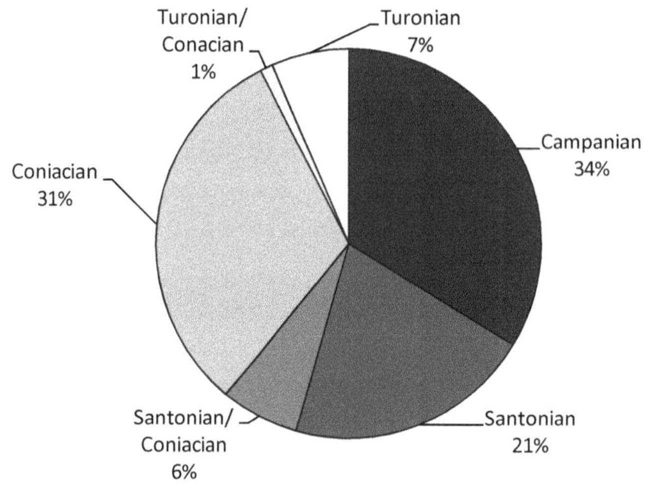


Fig. 9 : Distribution of the siliceous raw materials of the Cretaceous by stage.

| | nodule | slab | pebble | small bloc | total |
|---|---|---|---|---|---|
| Campanien | 25 | 5 | 1 | | 31 |
| Santonien | 17 | 2 | | | 19 |
| Coniacien | 27 | 2 | | | 29 |
| Turonien | 5 | | | 1 | 6 |
| Maastrischien | 2 | | | | 2 |
| total | 76 | 9 | 1 | 1 | 87 |

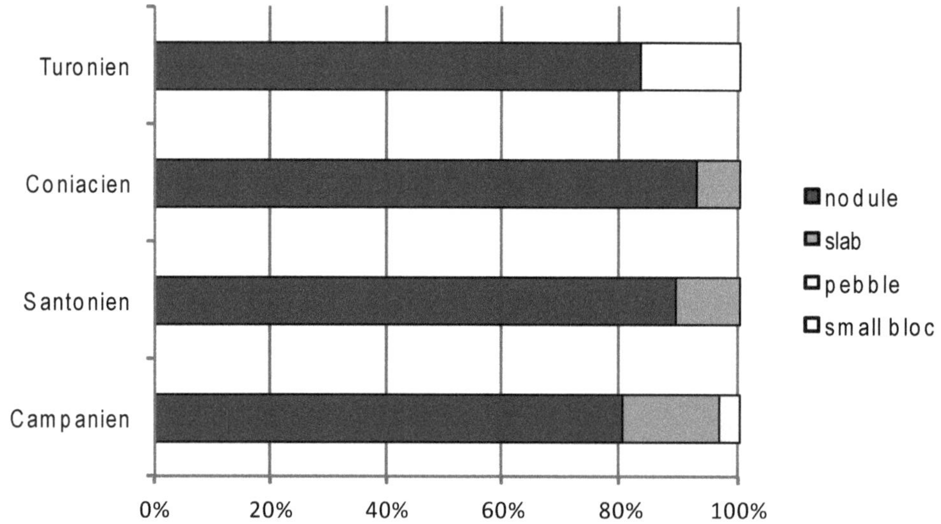


Fig. 10 : Morphology of the samples of the Cretaceous.

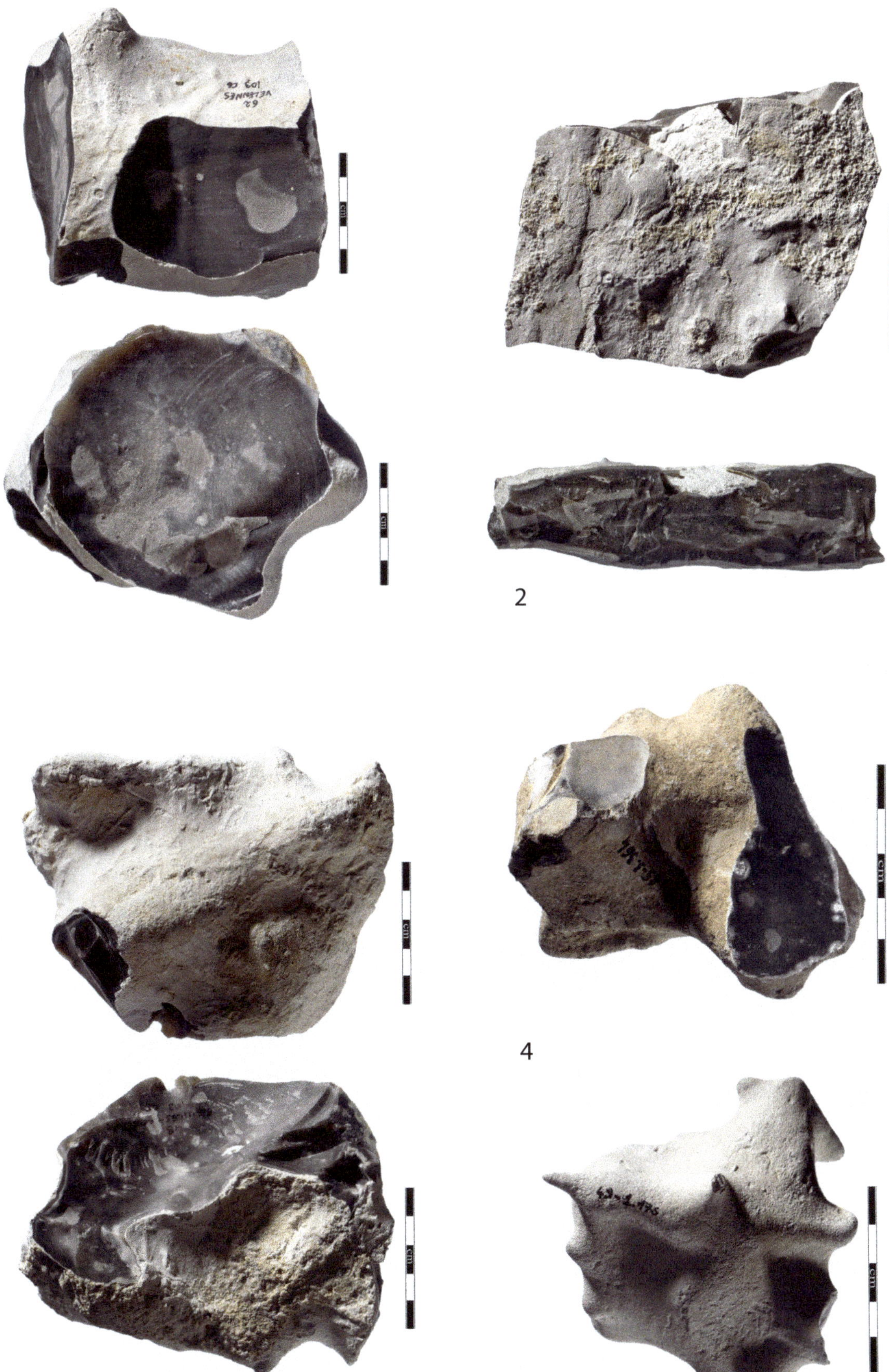


Fig. 11 : Silicifications of the Secondary. 1. Velennes, Campanien C6, 2. Hardivilliers, Campanien C6, 3. 3. Essuilkes, Santonien C5, 4. Remaucourt, Turonien C3b-c, 5. Villers-Outreaux, Coniacien C4-5 (photos S. Oboukhoff, CNRS).

| | pellicular | 1 - 2 mm | 2 - 5 mm | 5 - 10 mm | irregular | Total |
|---|---|---|---|---|---|---|
| Campanien | 28 | 3 | | | | 31 |
| Santonien | 12 | 4 | 1 | | 2 | 19 |
| Coniacien | 7 | 14 | 7 | 1 | | 29 |
| Turonien | 2 | 2 | 1 | 1 | | 6 |
| Maastrischien | 2 | | | | | 2 |
| | 51 | 23 | 9 | 2 | 2 | 87 |

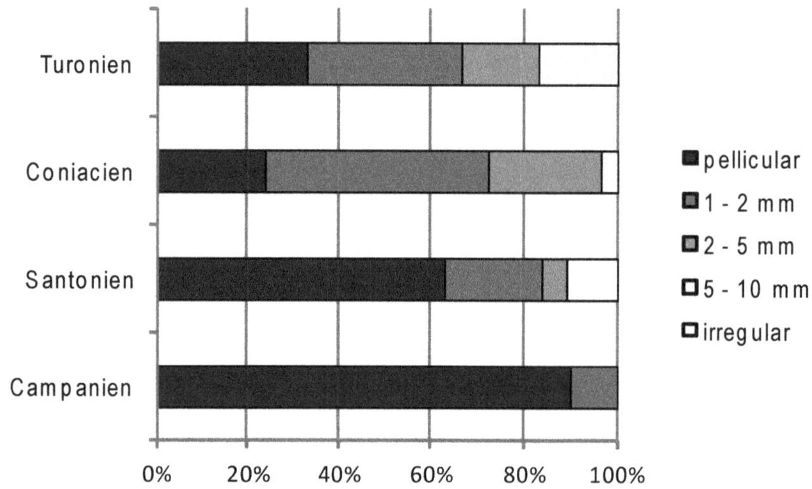


Fig. 12 : Cortex thickness of the samples of the Cretaceous by geological stage.

| | nodule | slab | small bloc | pebble | undet/flake | total |
|---|---|---|---|---|---|---|
| Muschelkalk | | 5 | | | | 5 |
| middle Jurassic | 1 | | 2 | 1 | 2 | 4 |
| superior Jurassic | 1 | | 2 | 1 | 3 | 7 |
| total | 2 | 5 | 4 | 2 | 5 | 16 |

Fig. 13 : Morphological characteristics of the samples of the Palaeozoic.

# Travels and transformations of flint: towards a new lithotheca model and dynamic map

Paul Fernandes, Anne Hauzeur, Pascal Tallet, Jean-Paul Raynal

## Abstract

The characteristics of their initial crystallization and the evidence for subsequent alterations present on the surfaces of prehistoric flint artefacts result from a range of interdependent physical, chemical and mechanical phenomena. Once decoded, the retrieved information allows us to assign a genetic/stratigraphic position to the flint as well as a post-genetic palaeo-geographical location (ajouter tes references ici, au moins le palevol sur les cortex et ta these). Both data sets reveal the story of an object and its successive residential contexts. This petro-archaeological analytical methodology is based on an optimization of the optical techniques commonly used in petrography, mineralogy, micro-palaeontology and morphoscopic examinations using a range of magnifications. The methodology provides a simple tool for classifying objects and identifying the sources of lithic raw materials. The limitations of the methodology are determined by the experience of the operator and the geographical spread of available reference collections.

**Keywords** : Siliceous rocks; Lithotheque; Reference collection; depositionnal process; southern France; GIS

## 1. Introduction

A study of the various alteration processes (chemical weathering) to which lithic raw materials have been subjected and which are recorded within their cortex allows connections to be made between the lithic objects themselves and the surface exposures from which they originate. These revealing characteristics allow us to see lithic objects from a perspective which connects them to landscapes and specific resource localities. The methodology used here also allows flint objects to be compared one to another within the contexts of time and space. Having determined these parameters, the genesis of each object can be described resulting in an understanding of where, why and how the lithics were collected during prehistory (Fernandes 2006, 2012; Fernandes and Raynal 2006; Fernandes *et al.* 2008).

A lithic reference collection has to integrate this concept by respecting all the steps that an object may encounter in its evolutionary chain from raw material to finished and discarded object. A simple inventory of lithic sources provides insufficient data for a complete analysis of the life of a lithic object. Expanding the parameters of the contexts from which it was recovered, used and discarded using meticulous descriptions of flint objects from within the same evolutionary sequences as indicated by the state of their surface characteristics, renders the lithic reference collection into a geographical data base that assists archaeologists to interpret the procurement and management of the lithic raw materials.

## 2. Historical Background

The lithic reference collection ("lithotheca") was built up in association with the production of a map that recorded the investigations conducted by several members of the collective research programme "*Réseau de Lithothèques en Rhône-Alpes*" (report 2007).[1]

The assembled data set shows the distribution of prehistoric lithics and the localities from which they were collected for two regions of central France; Rhone-Alps and Auvergne. The maps provide an overview of topographic locations, stratigraphic origins, secondary positioning, and altitude. During the first three years of the project the geographical information system (GIS) was totally devoted to registering the geological information accompanying samples stored in regional lithic reference collections. Despite the apparent faithfulness of the dots on the map to the data set, they failed to take into account the localised geological complexity of the areas over which the siliceous materials are distributed. Moreover, the scale of these initial maps was inadequate for extracting the detail which we were seeking. From 2009 onward a new GIS system, better adapted to our requirements and which enabled us to merge data from several sources has been used, (Guilbert 2000, Affolter 2010, Bressy *et al.* 2010).

## 3. Towards the new model

In 2011 our investigation and research area was extended to the whole of southern France in order to boost interregional connections between lithic objects and their source locations.[2] The northern limit is a more-or-less straight line between La Rochelle and Bourg-en-Bresse, which is the boundary of the area from which we have adequate data. Nevertheless, about one third of the area south of this line still needs further investigation. Besides the field information we gathered, the research programme has included information extracted from theses, published monographs, journal articles and reports that has been correlated with an analysis of 529 individual geological maps at scale 1:50,000 and their explanatory notes along with 252 data sheets from the website "InfoTerre" (http://infoterre.brgm.fr/) where data relating to drill-holes and their logs can be found.

The map resulting from the accumulation of all this data shows the topographical limits as well as the stratigraphical origin of the main flint formations in four areas: Auvergne, Rhone-Alps, Provence-Alps/Côte d'Azur and Aquitaine (Fig. 1). The detail to which each formation is mapped required the division of the main map into sections and these were made following the administrative boundary of the Departments concerned. This arrangement provided maps at a scale that allowed us to retain the required geological background information without any loss of detail.

The polygons on our maps define the primary flint outcrops as well as formations in secondary contexts while also indicating their interconnectedness, (fig. 2-3). The latter kinds of deposits are distinguished according to their stratigraphic and paleo-environmental origins. The resulting map is a dynamic portable document file (PDF) composed of three superimposed layers: the primary geological formations, alterites[3] and slope deposits or colluvium and lastly alluvial formations. The latter must be treated separately as alluvial transportation can assist pieces of flint to migrate from their original source location into another geological zone. A mouse-click on a formation on the electronic map opens a page containing descriptions, explanations and pictures at various scales (from that of the wider formation in the landscape to ultramicroscopic details). Zooming of these pictures is available at scales from 1:1,000,000 to 1:25,000.

The evolution of the landscape relief must be considered when mapping formations and this requires a geomorphological study of each secondary deposit that is identified and includes the evolution of waterways and various slopes within drainage basins. The types of raw material existing in the secondary formations may demonstrate the extremes for the genetic families of the lithics and the variations in deposition. Reconstructing the natural routes taken by each lithic object from its origin to its final resting place is an essential element of analytical petro-archaeology.

Toponyms for geological formations used on the maps are given as a term which localises precisely the named geological unit. It generally refers to the closest toponym that already exists on the base map (at 1:25,000 scale). For the large formations, the official toponym of the most important administrative unit inside the defined area has been retained. In exceptional cases and to simplify map use, the selected term takes into consideration the published terms already in use by archaeologists and geologists, e.g., "Bergeracois" for flint coming from the Late Campanian which is present in the alterites surrounding Bergerac.

Names of geological formations conform to those found on the international stratigraphic chart ICS G Ogg 2010 and colours chosen for the units we mapped generally conform to the schema used in the geological map of France at 1:50,000 scale (edition 2003). Variability within a single stage is distinguished by a raster, keeping the standard colour used for that stage. Flint formations have been drawn using "Adobe Illustrator" software with the geological formations vectorised from the geological maps at 1:50,000 scale.

Formation numbering is simplified, not listing the actual collection localities but rather the whole of a formation that includes flint of the same type(s). Thus, "formation" refers to an area where the lithics demonstrate common features and which therefore constitutes a total distinguishable whole. Once a formation is defined, a specific identifier is attributed to it that is composed of two numerals and a letter. The letter indicates the various geographical positions in which the same formation occurs. In the case of a scattered formation of the one type, the identifier series begins from the southeast occurrence. The inventory of each formation is accompanied by a brief description of the contexts and the flint type(s) recognised as occurring there. The descriptive notes are written by one author and reviewed by the others to obtain a consensus.[4]

Research protocol and recording has been consistent for each geological sheet we produced. Firstly, previously studied and already known deposits are mapped. Sometimes this data was available in the form of GIS maps of flint formations with the deposits marked using spot symbols. These spot locations were transferred to our geological base maps and used to generate polygons that defined the location and extent of the flint formations. Secondly, publications were searched for descriptions of formations that included flint and these too were transferred to our maps. Next, the dynamic legends on the "InfoTerre" website were perused because the research undertaken by the BRGM to vectorise the geological maps often provided relevant supplementary information that we could also include. It should be pointed out that our polygons may not include the occurrence of all flakeable raw materials in a particular area, either because nodules are not exposed on the surface or because those that can be seen may be of too small a size to be considered useable or they may be lithoclasts of non-human origin. Consequently, our maps provide a representation of the potential mineral resources and not necessarily only those that were exploited.

Clearly, flint objects found in an archaeological site are unlikely to be derived from a single given flint source if the surface traces of wear indicate that the object has been transported from or moved about within the catchment area where it naturally occurs. This is especially the case if surface features show that objects have been transported to locations well away from their zones of natural occurrence. It is unlikely that during Palaeolithic times the route from a raw material source to a habitation (and discard) site was a straight line joining the two locations - unless lithic use occurred very near its source. In any such case, the lithic object exhibits particular surficial modifications that lack signs of extensive mechanical or chemical modification. To more closely fit the reality of human environmental interactions, it was necessary to include a dynamic representation of source to discard locations. Each primary geological deposit is correlated with all surface formations for which it is the lithic source, and in our third generation maps, all flint outcrops with an identical lithological facies are grouped under the same label. In parallel, the notion of an evolutionary chain of modifications has been integrated into the data (Fernandes and Raynal 2006). Each type of flint is shown as originating from the primary deposit and proceeding towards the most remote secondary ones (Fig. 2).

The great majority of the lithic sources identified in our study are secondary ones that contain flint assemblages showing surface modifications that illustrate the manner in which they reached their present locations. Determining which surface characteristics correspond with the range of transport mechanisms requires considerable field time as geological maps rarely indicate the different flint facies that exist in surface formations and neither do they record the whole extent of the formations containing flint.

Our aim has been to identify and map the main primary and secondary occurrences of flint not as separate and distinct units but as an interconnected series, mainly because each individual deposit usually includes siliceous elements derived from one or more source deposits (Fig. 4). For each genetic lithic type it proves essential to map all the occurrences and reworked formations in which it occurs because each formation is just one step on the route that the object has taken from source to final discard and any one of the formations in which a particular type is found is a potential source deposit for that type when it is found in an archaeological context. The distributed range of any one genetic flint type can therefore be followed from its primary depositional contexts, through its sub-primary locations to its secondary occurrences and ultimately to its final resting places – and the same set of data is available on our maps for each flint type that has been identified. Our maps have had to take into consideration the different juxtaposed layers and various elevations of depositional occurrences which may be mixed or even superimposed one upon the other when multiple secondary deposits of the one type are involved. Despite these difficulties the maps show the likely route each lithic object has taken from its primary deposit to the most distant secondary one and the changes that have occurred on its surface during its journey.

After identifying the kinds of superficial markings present on a lithic object and its sub-surface characteristics it then becomes possible to allocate the archaeological artefact to a geological locality from which it originated. The degree of precision with which this can be done is far higher than can be achieved in the case of a classic study which does not take into account all the factors that our mapping project has considered and which correlate with the several characteristics identified on the surfaces of the lithic objects being examined. The protocols used in our laboratory analysis of objects enable us to identify the distinct geographical areas that were exploited within the broader occurrences of each flint type and the routes over which it travelled. This is because the most recent surficial alterations to the outermost layer of a lithic artefact are specific to the conditions to which it was subjected in its most recent geological location and characterise the last step of the "pre-deposit" phase of its existence just before it was collected for use by humans in prehistoric times.

The geological protocol can be summarised as:
- firstly, defining the position of the primary deposit,
- next, establishing the adjoining bordering zones of the primary occurrence where deposit variations can be observed,
- thirdly, delineating in this dispersal zone the different surface formations of flints with similar chemical weathering facies.

Secondary deposits are not considered as separate elements in our study, but rather as links in an evolutionary chain of modification that extends from the primary deposit out to the furthest occurrences. The first exposure of the potential resource, the mechanism that releases it from the bedrock and its further movement and transport through the environment are each capable of producing irreversible transformations that modify flint objects in specific ways. Taking all the identifiable variations into consideration is the key to determining with accuracy which particular individual deposits of flint were exploited during Prehistoric times.

## 4. The Lithic Reference Collection

A lithic reference collection must represent with accuracy the whole of the evolutionary chain of events to which a lithic object is subject. The totality of known modifications that have affected the one genetic type of flint must be grouped together; it is not sufficient to have just a simple stratigraphic description of the lithic resources available in a specific region. By archiving lithic samples using the categories determined by the state of their surface modifications that properly describes the evolutionary sequence through which they passed; the lithic reference collection becomes a vital geological/geographical repository that aids archaeological interpretations. Archaeologists can access, for any recorded genetic and deposit facies, comparative samples at a variety of scales (blocks, sometimes metallised fragments, thin blades, sampled sliced pieces etc.), illustrative of the successive transformations to which their materials have been subjected. The lithic reference collection should thus contain a full data set that will provide answers to an archaeologist's questions. No longer is the data set a simple accumulation of geological or geographical locations, it now contains accurate information about the dispersal areas specific to each genetic type of flint present in the studied region.

At the conclusion of this project, we can provide a strong diagnostic tool containing data on flint deposits for the south of France at a superregional scale and within which can be found the characteristics of genetic types of flint and examples of the processes to which they have been subjected. It has now become possible to better evaluate the regional exploitation of a lithic resource by comparing and contrasting the patterns of modification to which individual surface formations were subjected and the surface changes that are specific to each flint artefact.

## Acknowledgements
We thank Peter Bindon for the revision of English language.

## References

Affolter, J., 2010. La détermination des silex archéologiques: remarques méthodologiques. In *Silex et territoires préhistoriques. Avancées des recherches dans le midi de la France*. Les C@hiers de Géopré® 2010, electronic publication.

Bressy, C., 2002. Caractérisation et gestion du silex des sites mésolithiques et néolithiques du Nord-Ouest de l'arc alpin. Une approche pétrographique et géochimique, Thèse de doctorat, Université Aix-Marseille I, 677 p.

Fernandes, P. and Raynal, J.-P., 2006. Pétroarchéologie du silex: un retour aux sources. *Comptes rendus Palevol* 5/6, 829-837.

Fernandes, P., 2006. *Pétroarchéologie des matériaux siliceux utilisés au paléolithique moyen dans le sud du Massif central: méthodologie et résultats préliminaires.* Mémoire de diplôme de l'École des hautes études en sciences sociales. Toulouse: Écoles des hautes Études en sciences sociales, 183 p.

Fernandes, P., 2012. *Itinéraires et transformations du silex: une pétroarchéologie refondée, application au Paléolithique moyen.* PHD Thesis, Université Bordeaux I, ED Sciences et Environnement, Spécialité Préhistoire, Vol. 1 Text and figures 452 p., Vol. 2 Annexes, 164 p.

Fernandes, P., Raynal, J.-P. and Moncel, M.-H., 2008. Middle Palaeolithic raw material gathering territories and human mobility in the southern Massif central, France: first results from a petro-archaeological study on flint. *Journal of Archaeological Science,* 35, 2357-2370.

Guilbert, R., 2000. *Gestion des industries lithiques mésolithiques et néolithiques du Sud-Est de la France*, Thèse de 3ème Cycle, Université Paris-I, 369 p.

**Footnotes**

1 Affolter (Bauges, Bornes, Bugey, Jura, Middle Prealps, Savoy, Vercors) ; Morin and Cousseran (Buëch valley); Bressy (Bauges, Buëch, Bugey, Chartreuse, Chambaran, Diois, Isère valley, Royans, northern Vercors); Féblots-Augustin (Bugey, Ain); CAP Valence, Beeching and Brochier (Marsanne and Valdaine countries, Tricastin, Rhone valley, southern Baronnies); Bintz (Chartreuse, Diois, Isère, Vercors); Grünwald (Vercors, Royans, Isère, Diois, Chartreuse, Bugey); Fernandes (Central Massive, Cruas, Rochemaure, Saint-Montant, Saint-Marcel-d'Ardèche, Barjac-Issirac, Laval-Saint-Roman, Carsan, Vans, Aubenas, Crest, Taulignan) ; Masson (Jura, Savoy); Picavet (southern Vercors); Piboule (Auvergne, Loire, Cruas, Barjac-Issirac, Laval-Saint-Roman) ; Riche (Vassieux-en-Vercors); Vilain (Bugey, Savoy).

2 Collaborators are: Jean-Paul Raynal, Roger Séronie-Vivien, Michel Piboule, Alain Turq, André Morala, Jehanne Affolter, Pascal Foucher, Fréderic Bazile, Dominique and Françoise Millet, Vincent Delvigne, Marie-Hélène Moncel, René Liabeuf and Céline Bressy.

3 The French term 'alterite' is generally synonymous with saprolite, however, some authors use 'alterite' less specifically to refer to residual weathered material and 'isalterite' refers to saprolite.

4 The most complete description includes the following categories: local place, commune, formation stratigraphic code, author's name of the geological booklet, title and number of the geological map, collector's name and date, name of the person who has characterised the flint and date of the study, reference to the description and photo of the described sample (for ex. references for Rhône-Alpes are C. Bressy PhD (Bressy 2002, vol. 2, annexes 1, 3, 4, 5) and the J. Affolter articles (2010), bed-rock description (layer age and character, flint position and type, outcrop geographical location, etc.), description of the flint structure and texture, description of the flint mineralogy.

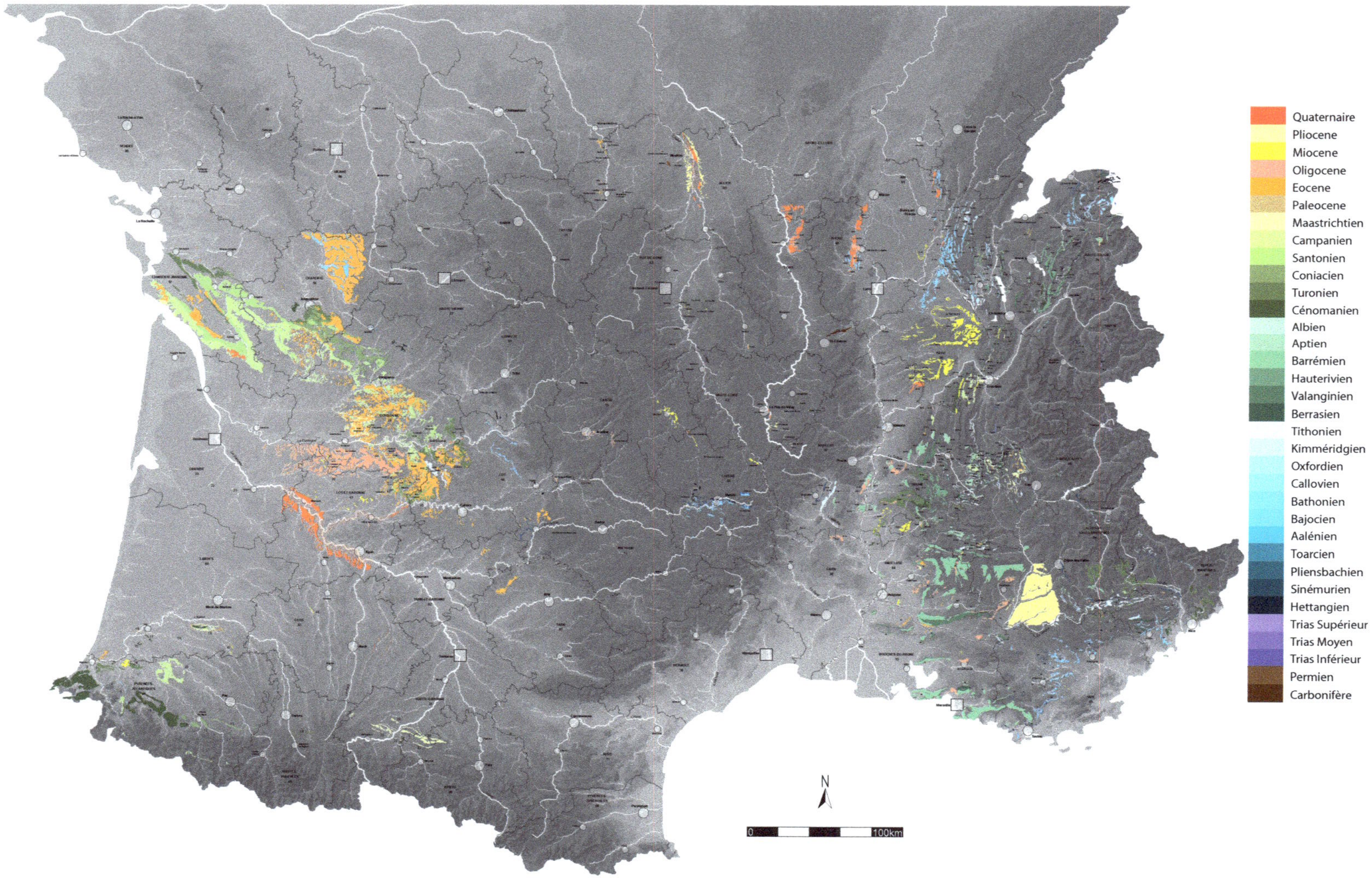


Fig. 1 : Flint formations in France. (CAD: P. Tallet). Authors: P. Fernandes, C.Tuffery, J.-P. Raynal, M. Piboule, M. and M.-R. Séronie-Vivien, A. Turq, A. Morala, P. Tallet, J. Affolter, D. and F. Millet, F. Bazile, P. Schmidt, P. Foucher, V. Delvigne, J. Liagre, S. Gaillot, A. Morin J.-F. Garnier, and M.-H. Moncel.

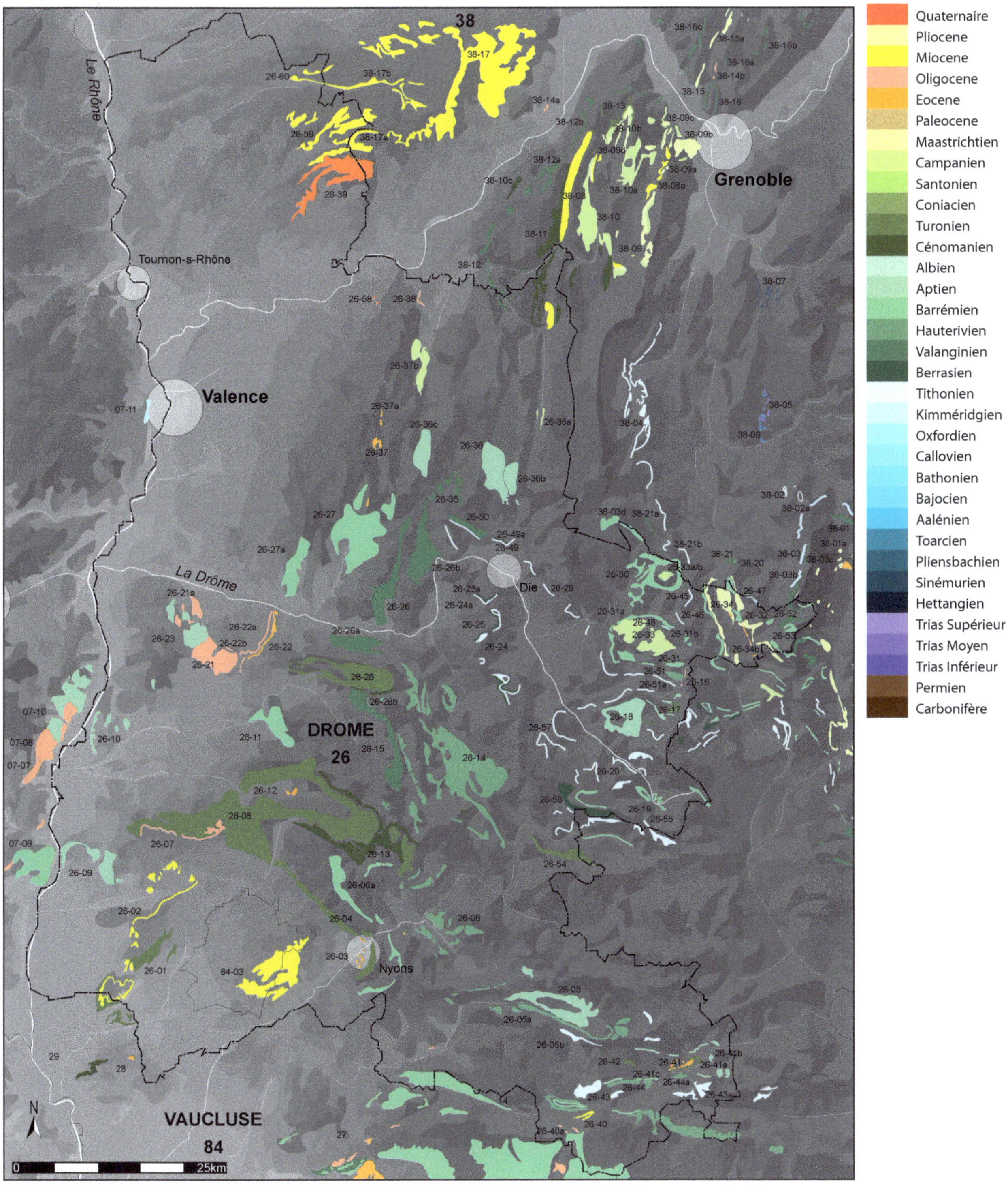


Fig. 2 : Flint formations of the department of Drôme. (DAO : P. Tallet). Authors: P. Fernandes, C.Tuffery, J.-P. Raynal, M. Piboule, M. and M.-R. Séronie-Vivien, A. Turq, A. Morala, P. Tallet, J. Affolter, D. and F. Millet, F. Bazile, P. Schmidt, P. Foucher, V. Delvigne, J. Liagre, S. Gaillot, A. Morin J.-F. Garnier and M.-H. Moncel.

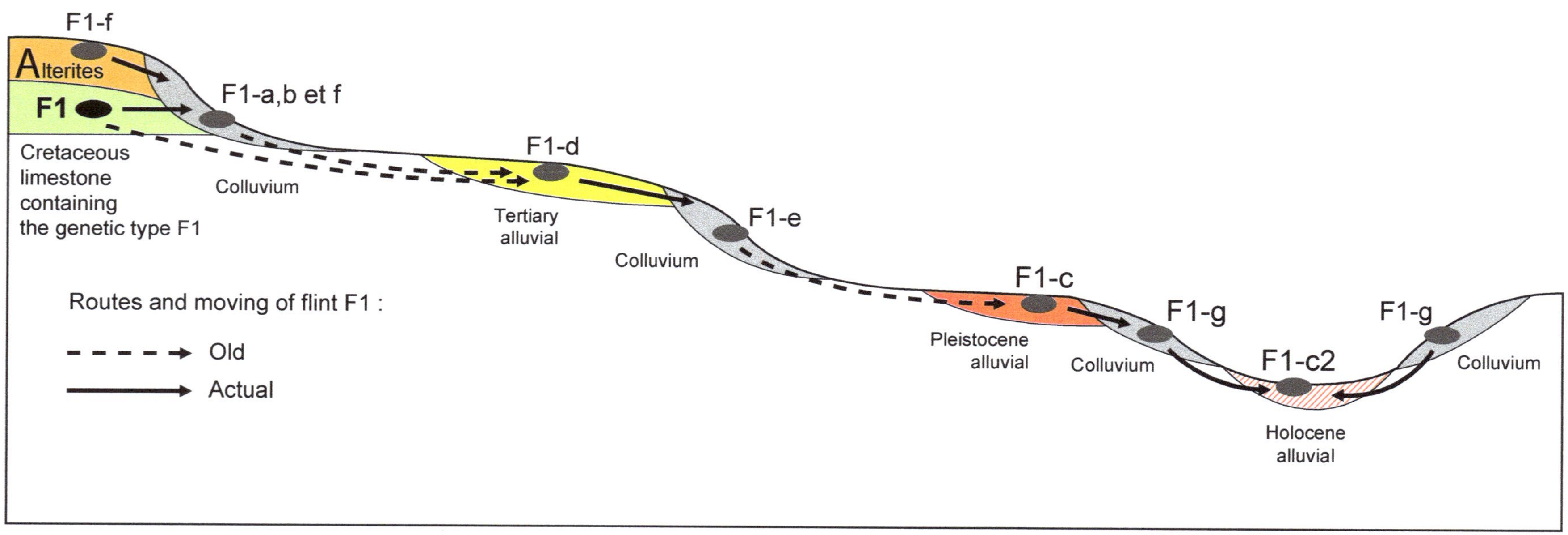


Fig. 3 : Dynamic representation of the evolutionary sequence through which a flint type passed from the primary deposit to its connected secondary deposits (CAD: P. Tallet). Each formation includes the same genetic type with a distinct weathering facies (mineral deposit facies): Each slope is an evolutionary geochemical unit. Therefore, although each block registered its own story each kind of environment resulted in a series of common features. These transformations are testimony to the surface formations types through which the flint travelled.

Fig.4 : Steps of lithic texture transformation. According to the distance from the primary occurrence to the final resting place, the general transformation of flint type during its journey is a mineral assemblages reorganisation and a progressive apparition of phases. Modifications are controlled at a macroscopic scale where figured elements are invariably disappearing.

# Investigation of new chipping floors at the flint mining site of Spiennes (Hainaut Province, Belgium)

Hélène COLLET, Michel WOODBURY and Jean-Philippe COLLIN

## Abstract

In 2011, the Service public de Wallonie with the help of the Société de Recherche préhistorique en Hainaut, conducted a seven month rescue excavation on plot 406b in Petit-Spiennes, at the location where a brand new visitor centre for the Neolithic flint mining site of Spiennes will be built in the near future. On this occasion, an area with chipping floors apparently perfectly preserved was brought to light. The investigation of such a feature, including thorough retrieval and spatial recording of the flint material is a first for Spiennes. It affords the opportunity to analyse the complete chaîne opératoire for the processing of bifacial tools such as axes, hachettes and flint picks as well as flake production. The notable absence of blade production from the entire investigated area disappointingly affords no opportunity to study this process so characteristic of the flint mining site of Spiennes. The preliminary results of this investigation and of the study of the lithic material are given here.

**Keywords:** Flint mining; chipping floors; bifacial production; axes; flint picks; Neolithic; Michelsberg; Hainaut.

## Introduction

The Neolithic flint mines of Spiennes are located in southern Belgium, in the Hainaut province, to the South of the Mons Basin. This flint-rich region, itself among a series of others such as northern France, Hesbaye and Duch Limburg, contains numerous mining sites of varying importance around the periphery of the basin: Spiennes, Baudour-Douvrain, Flénu, Ciply, Mesvin, Obourg and the newly discovered Villerot (fig. 1).

The Neolithic site of Spiennes is located south of the modern village on the slopes and plateau situated on both sides of the Trouille valley. Surface mining debris covers one hundred hectares. Such a concentration of flakes and other flint waste is related to the activity of knapping workshops. The area of extraction shafts is probably smaller. According to the current state of our research, this amounts to at least 50 hectares spread over three mining zones, the first at Camp-à-Cayaux, the second at Petit-Spiennes and the last near the Michelsberg enclosure (fig. 2).

Mining activities were carried on there from 4447/4329 to 2887/2704 cal BC, thus for more than 1500 years, and apparently without interruption. Two recent dates might even indicate exploitation up to 2279/2138 cal BC.

The site has been known since the second half of the 19th century and has been the subject of various investigations. These were carried out by scientific institutions as well as by private individuals. It was indeed not before the early 1990s that the site was finally protected and as such recognized as a monument and site. From the 1990s on, excavations have been regulated and monitored under the authority of the Public Services of Wallonia.

From 1997 to 2004, the Public Services of Wallonia and the Société de Recherche préhistorique en Hainaut have conducted archaeological campaigns at Petit-Spiennes, on plot 393c ahead of construction of a planned flint mines interpretation centre at this location (fig. 2, n°1). These researches allowed the thorough excavation of some flint extraction features dating from the beginning and the end of the 4th millennium BC, illustrating a true mining tradition, with mine shafts up to 10 m in depth. The project was eventually abandoned and it took several years before a new one emerged.

## 1. The Archaeological investigations carried out at Petit-Spiennes in 2011 and 2012

The project now nearing completion will be located above the mines excavated by the Société de Recherche préhistorique en Hainaut since 1953 on plot 406b (fig. 2, n°2). The building of this interpretation centre by the city of Mons has several objectives. Beside the exhibition area dedicated to the Neolithic and to the mines at Spiennes, it must make accessible to visitors the mines excavated by the Société de Recherche préhistorique en Hainaut, approximately 100 m² of underground workings at Petit-Spiennes. In addition it will accommodate further archaeological research.

Before construction it was possible to carry out an archaeological examination lasting seven months from March to October 2011. The cleaning of a 1500 m² area around the zone excavated by the Société de Recherche préhistorique en Hainaut revealed extensive archaeological remains dating from the Neolithic period, mainly flint extraction shafts and knapping areas.

The Neolithic remains were partially cut by a ditch 8 m wide crossing the excavated area from one side to the other. With the help of old maps, aerial photographs and earlier excavations, this ditch has been identified as part of the defensive lines of La Trouille. This defensive system was established in the late 17th and early 18th century by the armies of the French king Louis XIV to secure the northern border of his kingdom.

About sixty flint extraction features were discovered, all are underground workings. Chalk bedrock appears at a depth of 4.5 m in this area and the first three flint seams occur at depths between 5.8 and 8.5 m in the shafts previously investigated.

The surprise revealed by the investigation was the large quantity of flint spread at the base of the humic soil. On the basis of evidence gathered during the excavations, these spreads have been identified as sealed flint knapping floors dating to the Neolithic (fig. 4).
In 2012, thanks to a delay in the building project, a few extra square metres of floor 56 could be examined. The results of these last excavations are not presented in this paper.

## 2. Chipping floors from 2011

The chipping floors are mainly located in the north-west quarter of the investigated area. A few fragments of this dense flint layer also remain in the east and west (fig. 3). They are totally absent in the south, where the land suffered greater erosion.

This dense layer of knapping waste is located beneath a layer of humic soil level devoid of any artefacts, indicating a complete absence of plough damage to the chipping floor. The concentration of remains behind the 18th century ditch led us to hypothesize that the bank necessarily located behind the ditch, now gone, must have played a strong role in their preservation.
The cleaning and excavation of various areas destined to be destroyed ahead of the construction works have confirmed the presence of these remains over a very large area. These surveys also highlighted the stratigraphic complexity of what actually consists of successive concentrations of debitage interleaved with spreads of mining debris. These were recognized thanks to the presence in these layers of chalk as well as gravel from a Middle Pleistocene alluvial terrace (namely the Mesvin alluvial terrace) that occurs locally at a depth of 3.4 m in the substrate. The spoil from mining, such as loess, grey/blue sands and green sands, have however disappeared. These deposits lie on what we suspect to be the Holocene soil consisting of two thin horizons, a dark upper one above the second, lighter one, overlying the brown/orange clayey silt horizon.

Given the scale of the concentrations of debitage, only a part of the surface corresponding to chipping floor 56 was excavated. Five and a half month were devoted to its excavation. It is in fact the first time that so much time could be allocated to fully record a large portion of a knapping workshop at Spiennes. The excavated material will form a reference collection for Neolithic flint mining at Spiennes. What remained of the lithic layer was left in place and protected; it will be preserved under the building.

## 3. The knapping workshops previously excavated in Spiennes

This is not the first time that knapping workshop areas have been observed at Spiennes. While most of the land has been cultivated for a long time, various excavations have however revealed areas where workshops were preserved.

In 1965, François Hubert investigated a newly cultivated plot at Camp-à-Cayaux where the remains were in an exceptional state of preservation (fig. 2, n°6). This investigation allowed him to excavate four perfectly preserved workshops whose main activity consisted of the production of long blades. Axe shaping was another important activity there, alongside the manufacture of secondary products such as picks and flakes. Only a fraction of the lithics i.e tools, cores, roughouts and preforms was collected (Hubert 1969).

In 2005, chipping floors were brought to light at Camp-à-Cayaux during enabling works ahead of the building of a dwelling (fig. 2, n°3). In this case too, the plot was grassland which might never have been cultivated. As this property is located outside the heritage site, the intervention was short. Only three weeks could be spent on the examination of 150 m², affected by the future building. Given the time limits, it was decided to lift samples of the working floors in blocks (Woodbury and Collet 2007a). That these workshops were entirely given over to axe production was confirmed by the examination of a small sample of knapping waste. The latter shows the predominance of thin flakes of small size, mainly non-cortical (Collet et al, 2008). By contrast, the workshop does not yield any evidence of standardized blade production.

Other investigations at Spiennes must be mentioned. Of even smaller scale, these interventions once again showed evidence of well-preserved knapping activities close to the surface, offering an archaeological potential that should not be underestimated.

In 2004, the monitoring of the erection of a fence on plot 169a at the Camp-à-Cayaux (fig. 2, n°4) showed the presence of dense flint knapping levels in five postholes. Only a few axe roughouts were retrieved (Collet and Woodbury, 2007b).

In 2006, on the plot for the Centre for Archaeological Research (fig. 2, n°3), much disturbed by various installations related to the building such as drainage and tanks, some small spreads of debitage were discovered. Despite the limited size of the features preserved, there was strong evidence for standardized blade production (Collet and Woodbury, 2008).

## 4. Method

First of all, the knapping area was divided into metre squares. Initially, artefacts were retrieved one at a time and (according to thickness of the layer) by successive levels. However it soon became clear that this method was too slow, so it was decided to adopt a faster method, lifting the artefacts by 25 cm squares, but still by successive levels. This enabled the investigation of a significant area while maintaining an accurate three-dimensional record.

In addition, comprehensive soil samples were taken from a selection of the test squares covering the workshop. This was to enable subsequent granulometric studies. Sieving remains using a 2 mm mesh allows the quantification of small-size lithic waste. These analyses provide further confirmation of the in situ nature of the workshops. Study on experimental knapping remains tends to demonstrate that an in situ workshop should ideally deliver 60 to 80% of 2 to 4 mm chips regardless of knapping technique. Therefore, the over or under-representation of a granulometric fraction indicates sorting, which can be natural or anthropogenic (Bertran et al., 2006).

Pedological studies conducted by geologists from the Public Services of Wallonia accompanied the archaeological excavations. Longitudinal and transverse sections were cut one metre apart in order to record the deposition process. The sections dug beneath the workshop were carefully examined. Samples for micromorphological analysis were taken on this occasion as well as for palynological study. Such analyses should elucidate the nature of the soil under the workshop.

## 5. Preliminary Results

*The Lithic material from the area of the chipping floor 56*

The processing of this material is still in progress. So far the first three collecting levels have been cleaned, marked and sorted. The final statistics for the chipping floor material are not yet complete; the flint waste has not yet been studied. Nevertheless the examples of lithic artefacts singled out on a first examination of the material form a representative sample of the *chaînes opératoires* represented here.

The area of chipping floor 56 was mainly given over to the production of bifacial tools. The most advanced of these show that axes were the intended end products. Second in importance was the production of hachettes, chisels and flint picks. Given the wide range of the products, some bifacial roughouts remain after examination in the 'indeterminate' category. These are examples specimens that cannot be assigned to a specific category either because the object is fragmentary or clumsy shaping masks the intended end product.

Several complete roughouts discarded at the earliest stage of shaping show that the initial blanks measure from 180 to 210 mm. The smallest axe roughouts discarded measure from 140 to 150 mm. The hachette preforms are even shorter: from 105 to 130 mm.

Various blanks were selected. Knappers chose natural nodules, fragments of nodules (Fig. 5, n° 1) or even large flakes (Fig. 5, n° 2). The intention was to select an appropriate blank suitable for an axe preform. For that reason oblong and flat nodules were of particular interest as well as large cortical flakes. Thanks to the convexity of their upper face, the latter required little preparation in comparison with nodules. In some cases knappers selected broad and short flakes to make axes. The striking platform of such flakes is thus located near the middle of one of the two side edges and its removal was a difficult operation in the manufacturing process. This led to the abandonment of three specimens where the platform could not be removed.

When thick blanks were used thickness was an additional problem to be addressed. In some cases the knappers resolved this by taking off a long flake parallel to the long axis of the roughout resulting in a channel-shaped removal (Fig. 6, n° 2). This type of preparation is fairly uncommon but was observed on four fragmentary bifacial preforms.

Among the axe roughouts, few pieces were encountered at an advanced stage of shaping. In most cases only the early stages of preparation were identified, that is when the general morphology of the roughout is achieved by means of large removals usually made with a hard hammerstone. This initial stage produced a fairly regular preform resembling rectangular parallelepiped. This was arrived at by various means. The preform could be reduced bifacially from alternate edges, bilaterally but from one face only, or bifacially and bilaterally. Only two broken axe butts at a more advanced stage of preparation were recovered from chipping floor 56. Here, shaping was finished using a soft hammer, evening out the ridges (Fig. 7, n° 1).

Without a more comprehensive study of the waste flakes from the chipping floor it is not possible to explain the presence/absence of the various types of artefact. For instance, we cannot tell whether only the critical stages of preparation, namely the initial and the general shaping of preforms were carried out on the chipping floor before these were taken away and finished in another place or whether these initial processes carried a high rate of failure while further stages were more successfully achieved, leading to their low frequencies in the discarded material.

The majority of preforms coming from the chipping floor show transverse fractures which occurred during the manufacturing process (Fig. 6). 52 preform fragments were recovered for only seventeen complete ones. In one third of these cases, the fractures cannot be attributed to the knapper. They occurred because of inherent flaws in the raw material. On these broken fragments knapping errors such as deep or hinged removals are infrequent while the shaping is more advanced than on the complete ones. For the complete roughouts, only three out of the seventeen specimens were discarded because of defects in the raw material. In most cases they were abandoned because of volume management problems which appeared early in the shaping process, compromising bifacial and bilateral symmetries, like the deep removals, failure to reduce the thickness on one edge or else a problematic distribution of these removals. Just one preform, a complete roughout, shows no visible defect. This piece could have been discarded because of its small size: only 130 mm long.

Although numerous roughouts were broken during manufacture, examination of these reveals a high level of knapping skill hampered by a relatively difficult raw material.

The hachette roughouts form a distinct category. These pieces measure from 105 to 130 mm. This group is mainly composed of complete specimens. Only two out of seventeen are broken. With these smaller artefacts, no quality problems with the raw material were encountered. Neither were they discarded because of their size. The major causes of abandonment were problems encountered during knapping. The specimens frequently show deep removals and/or hinge fractures (Fig. 7, n° 2 and n° 3). These last are particularly numerous and as they impeded thinning, they were a major cause of discard. In this category knapping skill is less visible. Two roughouts showed particularly clumsy shaping.

Interestingly though, these hachette preforms reveal a more advanced stage of preparation. Two thirds of them were discarded during the finishing stage. At some stage the flint knappers found themselves unable to proceed: this might be after successfully shaping the cutting edge, or the whole of the distal portion, or even both cutting edge and one side. One example shows once again the problem described in axe manufacture: the striking platform from which the blank was struck was located in the middle of one side edge and proved resistant to reduction (Fig. 7, n° 2). Hachettes are likely to be a secondary product as is shown by a fragment of large broken roughout reworked into a hachette.

The chipping floor area yielded numerous flint picks, a mining tool typical of Spiennes, which might appear at first sight surprising. These rather large tools are knapped from two (bifacial picks) or three edges (triangular-sectioned picks). The working edge (the tip) is formed by the meeting of the two or three faces thus produced. The flint picks recovered measure from 145 to 207 mm. That these tools are opportunistic is shown by the selection of suitably-shaped blanks. For instance flakes or nodule fragments with a natural trihedral tip were chosen for easy conversion into a working edge. In two cases discarded axe preforms were recycled as flint picks. This has been observed before during analysis of picks from an excavated feature (Collin and Collet 2011). Most of the flint picks coming from the workshop are preforms which were shaped in situ but were not then used. Three were broken during manufacture. Eight are not finished tools but still preforms. Some are barely shaped. Others display worked edges but an unfinished tip. Finally three are fully shaped but were left on the workshop floor. However, a closer look at these specimens reveals some problems. In one case the flint pick was manufactured on a twisted flake. In another the working tip is too fragile because of a flaw in the raw material. In a third case the working edge is markedly off-centre. Lastly, one example has one possibly unfinished edge. Five flint picks bearing traces of use were found on the chipping floor. Two were found on the surface. The three others clearly belonged to the workshop and two of them were even used as hammerstones.

Flake production is well documented in the workshop area. 40 flake cores were recovered as well as 'technical' pieces like core rejuvenation flakes. The original locus of this activity as well as its purpose in the context of chipping floor 56 must be further investigated. On the other hand, the absence of any significant blade production is notable even though Spiennes is famous for its long blade production. During the present intervention only a few elements were recovered: one partially cortical blade, one blade fragment characteristic of the mining production but still unretouched and a hachette preform made on an exhausted blade core fragment (Fig. 7, n° 3). The latter retained a small portion of the original flaking surface, and also of the side and back. It is absolutely typical of the standardized blade production at Spiennes.

Several tools were discovered. Most of them seem to be rather opportunistic. They are poorly retouched flakes removed from various kinds of blank. Nevertheless two end scrapers, two tranchets, one burin and three arrowheads were found. A bone tool was also discovered in the area of chipping floor 56. Though damaged by the acidic soil, its morphology is still visible. It is made from a long bone which has been split longitudinally. Its end has been rounded into a spatulate form. Several similar examples are known at the flint mining site of Mesvin "Sans Pareil", a few kilometers from Spiennes, where it was considered to be a mining tool (Hauzeur 2011).

Knapping tools are represented by sandstone hammers as well as flint examples bearing percussion marks. Among the 163 sandstone artifacts recovered, only two are complete hammerstones, weighing respectively 465 g and 559 g. 35 fragments were collected with weights ranging from 14 to 984 g. The rest are small flakes. The polished surfaces of two small fragments reveal that some hammerstones were reused polissoir fragments. To date 20 flint examples with percussion marks have been identified but further careful examination of the material will probably bring more to light. They are mainly hammerstones but other percussive uses may yet be distinguished. Generally, they are products of the recycling of discarded bifacial tools. They weight from 358 to 687 g.

### *Feature in relationship with the Chipping floor*

Near the bottom of the chipping floor, between two layers of flint waste, a hearth was discovered at the sixth flint collecting level (Fig. 3). Few slight fire marks revealed its presence as well as material in its vicinity. Numerous little pieces of burnt bones were spread over square meters around. Charcoal fragments are rare. Various potsherds were found in the same area together with two of the three arrowheads coming from the chipping floor. The hearth was covered by an important flint waste showing that activity went on afterwards for a long over the chipping floor 56.

### *Chronology*

Type-fossils recovered during the excavation of the workshop include three arrowheads from different locations in the chipping floor (fig. 8) as well as potsherds. These triangular arrowheads with edge retouch are common on Belgium's Michelsberg sites dating to the Middle Neolithic (4300-3700 cal BC). The potsherds, with their characteristic fabric tempered with crushed flint, are also typical of Michelsberg pottery from the Scheldt basin. Flint pick production also fits perfectly with a Michelsberg date. In fact this is the only mining tool used in the extraction features that could be assigned to this period thanks to radiocarbon dating, such as the deep shafts of Camp-à-Cayaux, shaft 53.2 on plot 406b and shaft 20 on plot 393c (both at Petit-Spiennes) (Toussaint *et al.* 2010).

This initial dating needs to be confirmed by radiocarbon dating and detailed stratigraphic analysis of the deposit in order to check out the possible presence of later activity on chipping floor 56. Indeed the site of Spiennes was exploited for more than 1500 years and extraction features were still being dug during the third Millennium.

## 6. Perspectives

The study of chipping floor 56 undertaken in 2011 and 2012 is just beginning to bear fruit. It already allows a better understanding of bifacial flint production at Spiennes. Numerous analyses, such as the examination of flakes, have yet to be pursued. Study of the archaeology itself will be furthered with the help of the granulometric, spatial and stratigraphic analysis of deposits.

## References

Bertran, P., Claud, E., Detrain, L., Lenoble, A., Masson, B. and Vallin, L., 2006. Composition granulométrique des assemblages lithiques, application à l'étude taphonomique des sites paléolithiques. *Paléo* 18, 7-36.

Collet, H., Hauzeur, A. and Lech, J., 2008. The prehistoric flint mining complex at Spiennes (Belgium) on the occasion of its discovery 140 years ago In P. Allard, F. Bostyn, F. Giligny and J. Lech (eds.), *Flint mining in Prehistoric Europe: Interpreting the archaeological records.* European Association of Archaeologists, 12th Annual Meeting, Cracow, Poland, 19th-24th September 2006 (BAR International Series 1891), 41-77.

Collet, H. and Woodbury, M., 2007a. Mons/Spiennes : découverte d'ateliers et de puits d'extraction du silex rue d'Harmignies. *Chronique de l'Archéologie wallonne* 14, 39-40.

Collet, H. and Woodbury, M., 2007b. Mons/Spiennes : indices miniers néolithiques dans la parcelle 169[a] au « Camp-à-Cayaux ». *Chronique de l'Archéologie wallonne* 14, 45-46.

Collet, H. and Woodbury, M., 2008. Mons/Spiennes : fouille préventive des abords du Centre de Recherches archéologiques, anciennement appelé musée du « Champ-à-Cailloux ». *Chronique de l'Archéologie wallonne* 15, 35-37.

Collin, J.-Ph. and Collet H., 2011. Mode d'acquisition, stigmates d'utilisation et motifs d'abandon des outils d'extraction de la " ST 20 " de Petit-Spiennes, Spiennes (Hainaut, Belgique). *Anthropologie et Préhistoire* 122, 1-56.

Gosselin, F., 1986. Un site d'exploitation du silex à Spiennes (Hainaut), au lieu-dit "Petit-Spiennes", *Vie archéologique* 22, 33-160.

Hauzeur, A., 2011. Extraction et terrassement à Mesvin – « Sans Pareil » (Hainaut, Belgique). Outils miniers en matière dure animale de la collection Lemonnier (IRScNB), *Anthroplogica et Praehistorica* 122, 87-106.

Hubert, F., 1969. Fouilles au site minier néolithique de Spiennes. Campagne de 1965. (*Archaeologia Belgica* 111), Brussels: Service national des Fouilles.

Toussaint, M., Collet, H. and Jadin, I., 2010. Datations radiocarbones d'ossements humains du site minier néolithique de Spiennes (Mons, Hainaut), première approche. *Notae Praehistoricae* 30, 73-80.

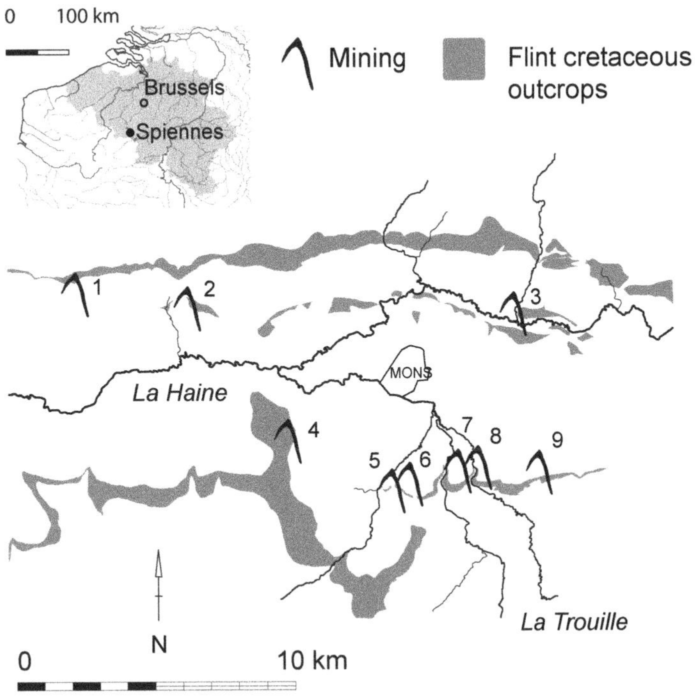


Fig. 1 : Spiennes and the flint mining sites of the Mons basin. 1- Villerot, 2 – Baudour-Douvrain, 3 – Obourg, 4 – Flénu, 5 – Ciply, 6 – Mesvin, 7 – Spiennes 'Petit-Spiennes', 8 – Spiennes 'Camp-à-Cayaux' (graphics M. Woodbury, SPW).

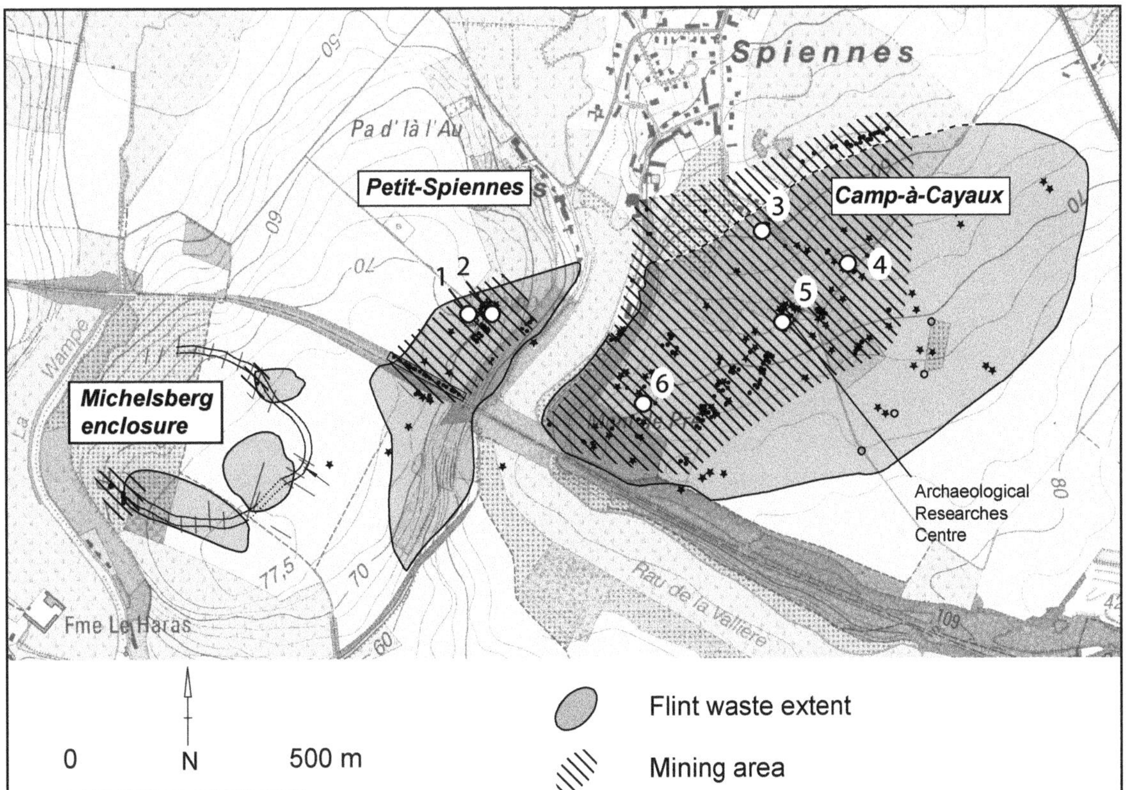


Fig. 2 : The flint mining site of Spiennes and the excavations mentioned in the paper. 1 – 1997-2004 excavation on plot 393c, 2 – 2011-2012 excavation on plots 406b and 406c, area of the shafts excavated by SRPH, 3 – 2005 excavation on plot 51c, 4 – 2005 excavation on plot 169a, 5 - 2005 excavation on plot 33h and area of deep shafts, 6 – 1965 excavation on plot 37 by François Hubert (graphics M. Woodbury, SPW).

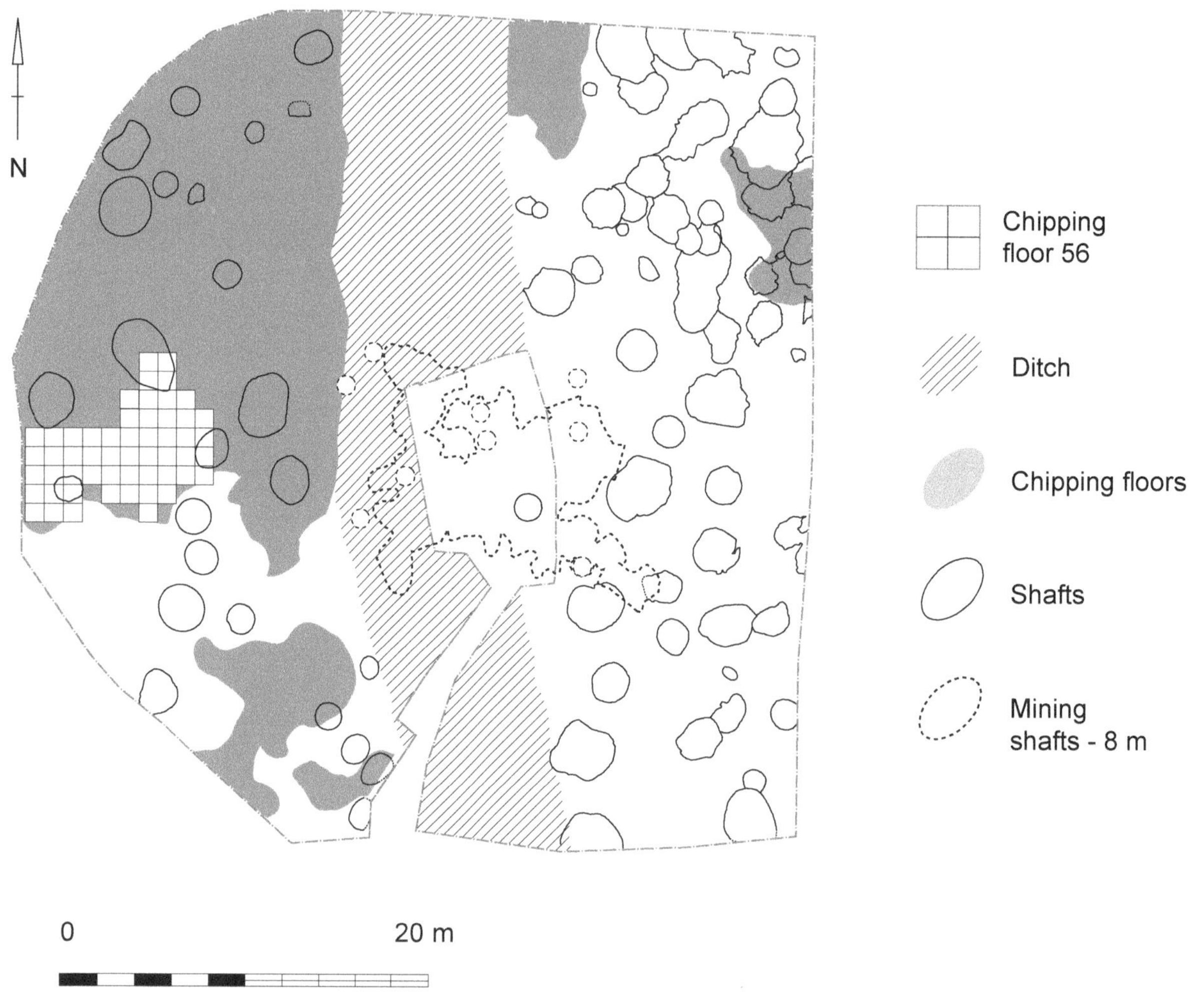


Fig. 3 : Map of the area excavated in 2011-2012 on plots 406b and 406c (graphics M. Woodbury, SPW).

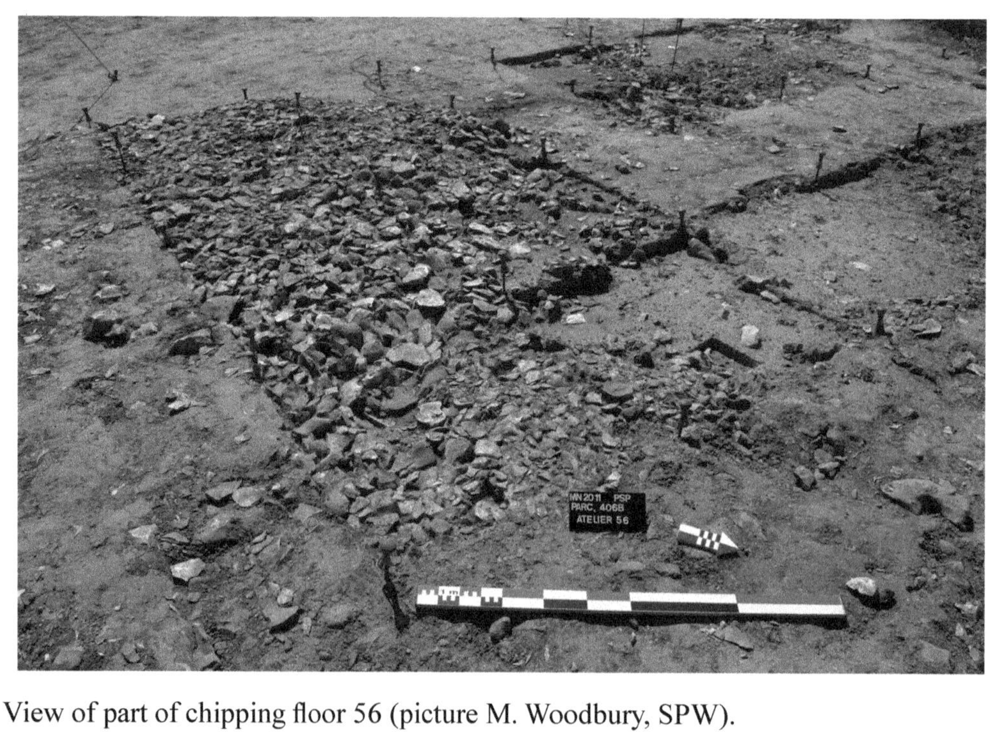

Fig. 4 : View of part of chipping floor 56 (picture M. Woodbury, SPW).

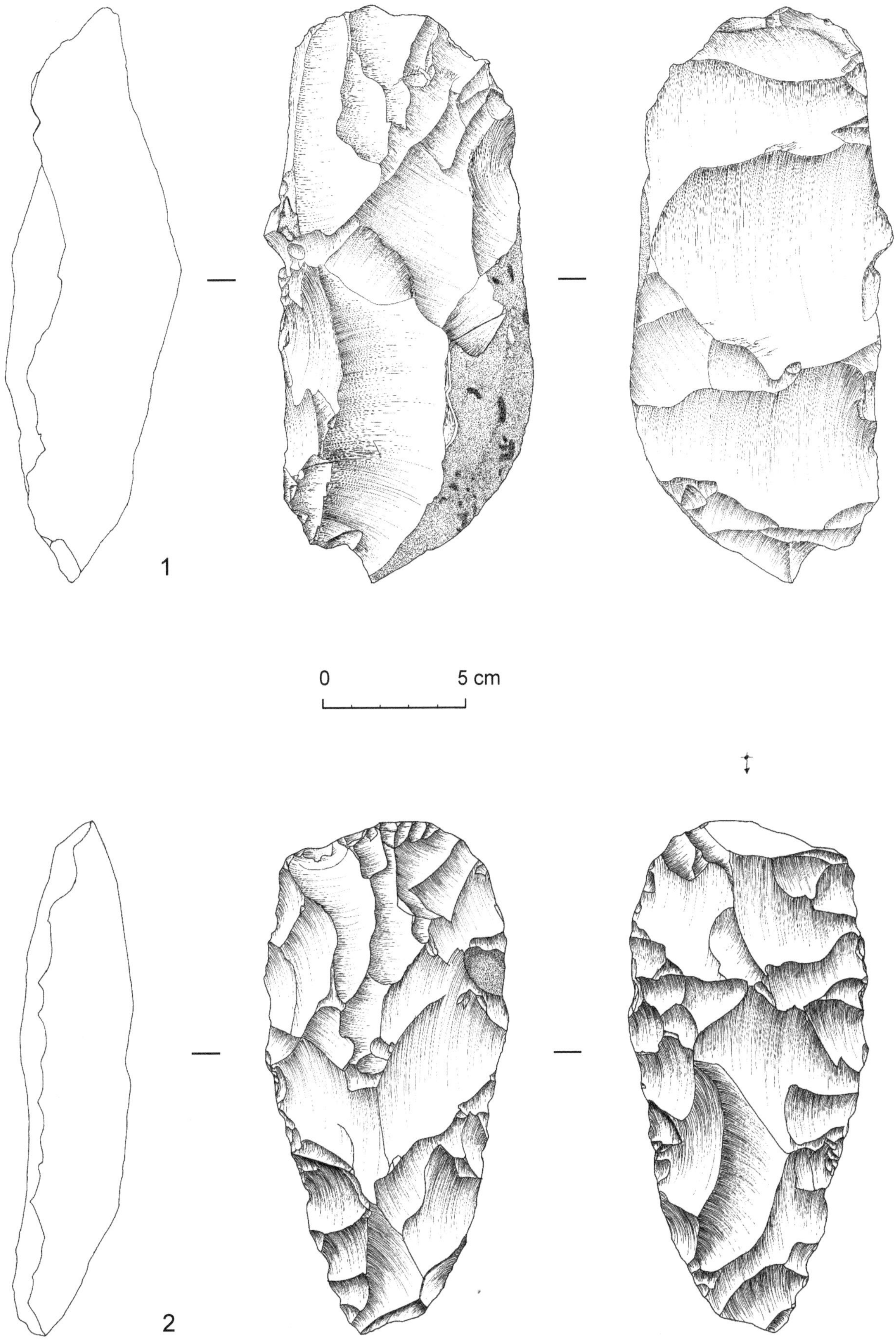


Fig. 5 : 1. axe roughout at an early stage of preparation, 2. axe preform (drawings A. Rosart, SRPH).

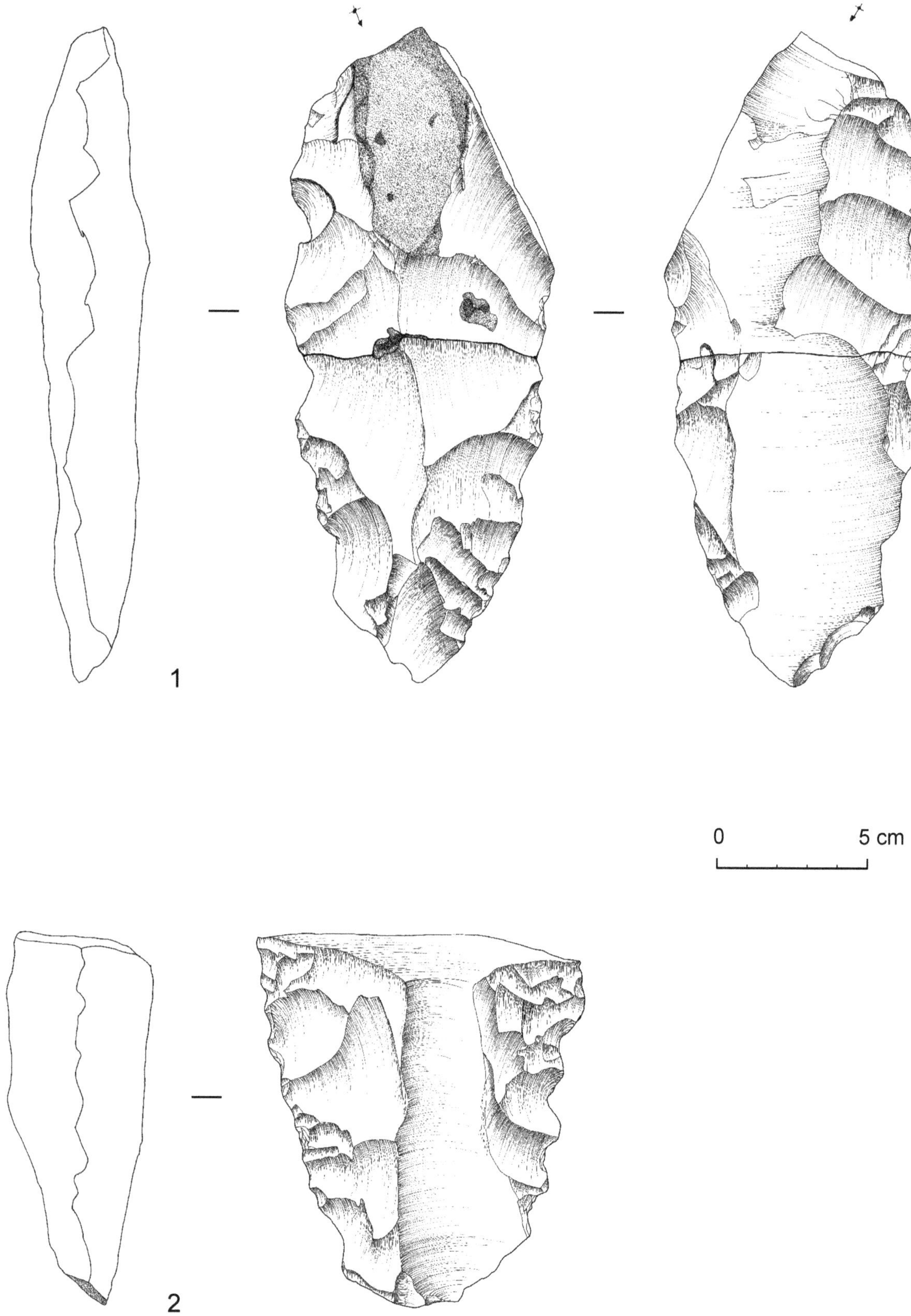


Fig. 6 : Axe roughout and axe preform fragment showing transverse fracture (drawings A. Rosart, SRPH).

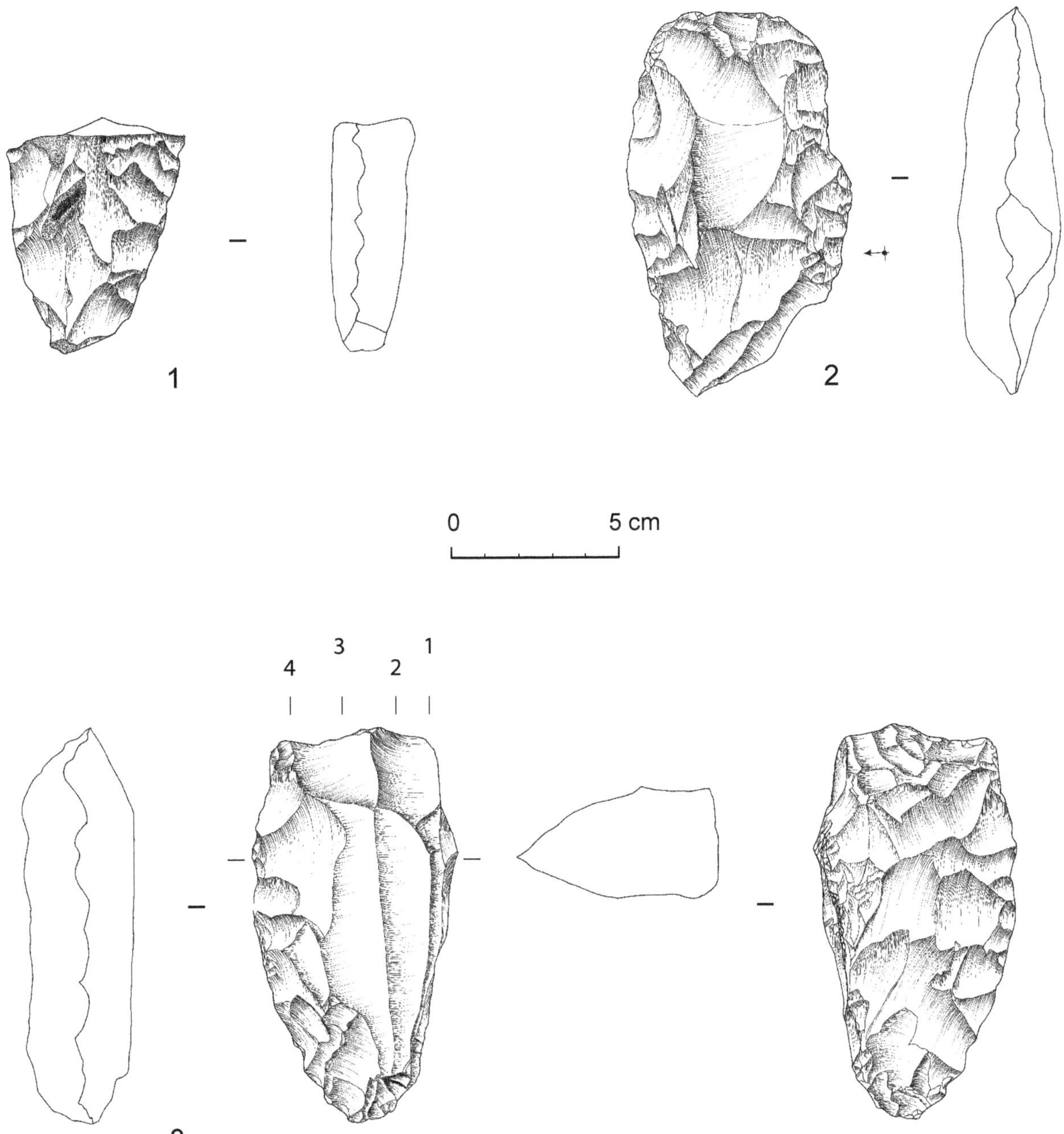


Fig. 7 : 1. axe preform fragment at an advanced stage of preparation; 2-3. hachette preforms (drawings A. Rosart, SRPH).

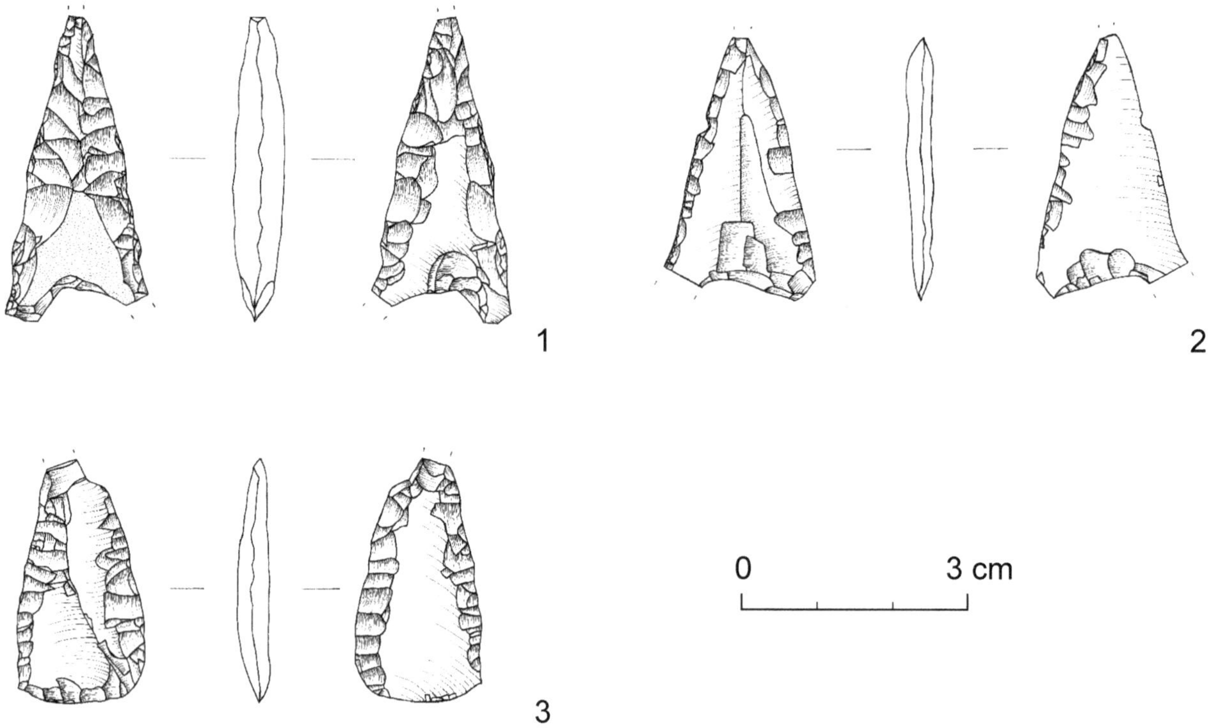


Fig. 8 : Arrowheads from chipping floor 56 (drawings A. Rosart, SRPH).

# Neolithic flint axes in the Loir Valley (France): quarrying, production, and distribution

Harold Lethrosne, Jean-Marc Lecoeuvre


## Abstract

Since the end of the nineteenth century several flint axes production workshops have been located in the Loir valley. The presence of these workshops is revealed by the presence of numerous sizeable wastes, among them many axes roughout discarded during fabrication. We have follow through the examination of many of these roughouts which were gathered from the surface originating from these workshops and a sixty kilometres radius around them as all those produced in the same flint stones. A technological study twined with a space repartition analysis of the products allowed us to draw out the surroundings of a first step diffusion of those flint axes in the Loir Valley. It seemed then that the Loir valley played a main part as an axe of circulation and diffusion of those preformed tools in the local Neolithic.


**Keywords:** Flint mine; axe production; diffusion; Loir valley; Neolithic.

## 1. Geographical and geological setting

Our geographical area of research encompass part of the Paris basin. This sedimentary basin allows a versatile use of flint stone exploitation during prehistory. During the Neolithic era, many mine devoted to flint extraction are founded. More precisely our area of study is located in the south-western part of the Paris basin, focused on the Loir valley, tributary of the Loire, to the north of the Loir-et-Cher department close to the town of Vendôme (Fig. 1).

The Loir valley separates two distinct geological entities. To the north-west, le Perche with chalks from the cretaceous Mesozoic era and to the south-east, the flint Beauce plateau from the Cainozoic era. In this area of the Loir Valley, the path of various tributaries and dry valleys follow accidents between two Cretaceous compartments. Their incision merges together two sedimentary geological formations from the Upper Cretaceous, especially the flint clays of the Campanian covering the flint chalks from the Turonian-Senonian (Manivit and *al.* 1983). In these valleys, flint nodules of Secondary flints outcrop.

## 2. History of research

Since the end of the 19th century to the beginning of the 20th century, many surface gathering of flint pieces have occurred around Vendôme (Maricourt, 1872 and St-Venant, 1917 et 1918). Those gatherings have been perpetrated by members of the « Société Archéologique, Historique et Littéraire du Vendômois ». Those precursors of local prehistory have found in this area the roughouts of bifacial and stone axes, alongside a great amount of flakes. Hinting them to the presence of flint blade axes' workshops. Instantaneously, they associated their theory with the occurrence of many polishing stone found in the Loir Valley and along its tributaries.

However there won't be any confirmation of those workshops until the end of the Sixties. After the digging of a drain ditch for agricultural purpose, following the path of a small dry valley, a large amount of flint cutting wastes were found (flakes, bifacial roughouts, pieces of flint nodules, etc.). This lead to a first campaign of digs between 1969 and 1973 lead by Jackie Despriée. The camp is situated on the town of Pezou at the lieu-dit « La Chenevière-Dieu », which is located on top of the southern versant of the Loir Valley. The digging area is located on the western edge of a small valley whose thalweg arbours an intermittent stream tributary to the Loir River. The totality of the excavation covers only thirty square meters. The gathered results showed that the trench dug for draining crossed a flint mine along side piles of flint axes production (Despriée 1986 and 1997). The recent reprisal of the site study especially of its flint artefacts gave us proofs for a highly specialize extraction site for this material (Despriée and Lethrosne tbp) and the production of axe blades during the recently Neolithic (Lethrosne 2012).

## 3. Data and method

Our acquaintance with the flint mines of the Loir Valley are not all on the same level of knowledge. The data is composed of: a digging expedition located on a small site at Pezou « La Chenevière-Dieu, old bibliographical notes of the axes production sites and flint axe blades roughouts from surface gathering therefore not archaeological. In order to better understand those mines in the south-western part of the Paris Basin we decided to ask ourselves a few problems. First and foremost we tried to find the different sites specialized in material extraction and/or specialised in flint axes production. We then decided to work on the characterisation of those flint axes production chains. And finally we ended up analysing the space repartition of the produces from the early phases of the operating chain, on a roughly measuring sixty kilometres radius, shaping out the outline of the Neolithic mines network of the mid Loir Valley.

We started by comparing various older bibliographical notes on the existence of workshops specialised to axes production, with the concentrations of axes roughouts found during surface gatherings, because such concentrations tend to indicate the localisation of a production workshop. We then tried to identify and cartography the totality of the sites linked to those activities and worked on the blade axes production *chaîne opératoire*., where the mineral isn't perfectly mineralized.

We created a database with a number of 395 flint bifacial pieces found by gathering and kept in local museum or on private properties. We focused on flint stones originating from the Campanian clays nodules. This material outcrops in this part of the Loir Valley where it was used during the Neolithic for axes production. It is a Campanian flint (upper Cretaceous) in the shape of big nodes, sometimes branched, with a relatively thin cortex of off-white color. The flint stone is grey beige in color in its cortical area whereas its core is blue-grey. The material contains numerous mineral as well as fossils, which sometimes are a centimetre or more in size, there also is important area where the mineral isn't perfectly mineralized, making the flint stone of mediocre quality for knapping.

The first step of the study was on the production of the *chaîne opératoire*, we took into account every products belonging to the different steps of production including polishing. The end products, polished axes weren't taken in to account, because previous studies have shown that this last step was disconnected from the main production both in space and time. We studied the very first steps of the *chaîne opératoire* from the bifacial prepararation to the blade regulation step when the blade is ready to be polished. We used the classification made by F. Bostyn for the technological analysis of stone furniture of Jablines « Le Haut Château » flint mine (Bostyn, Lanchon, 1992).

This technological classification is composed of three main categories:

- Bifacial preparation: this category encompasses the totality of the pieces having gone through all step of the preparation from the unilateral, monofacial shaping to the bifacial outer trim shaping. Large cortical areas may remain on one or both faces, if nodules or nodule pieces have been for the shaping out.

- Roughout: This particular step concerns pieces whose frame show off an advanced bifacial shaping. The cortical surface is thined out, a sharp face is drawn out but, as for the edge it is mostly sinuous.

- Axe: It's the last category pertaining to the shaping out of the *chaîne opératoire*. The piece arbour a homogenisation of its edges and of its cutting edge by a habile work of sharp retouches. The cortical part disappears or become irrelevant, the ridge is straight, and as such those are ready to be polished.

For the purpose of our study we have taken into accounts both those complete and those partially fragmented found from surface gathering. We identified 93 bifacial preparations, 258 roughouts, and, 44 axes (Table 1).

| | whole | fragment |
|---|---|---|
| Bifacial preparations | 58 | 35 |
| Roughouts | 159 | 99 |
| Axes | 40 | 4 |

Tabl. 1 : Occurrence of the analysed products.

The cartography of the data among the three categories of the *chaîne opératoire* allowed the study of their diffusion on a regional scale. This work will be followed through by the analysis of the finished products: the polished axes, in order to better understand the *chaîne opératoire* but also to allow the study of their diffusions.

## 4. Flint mines and workshops of axes' production

The various origins of data (bibliographical and archaeological) allowed us to find 10 locations devoted to the production of flint axes. These sites are located on both banks of the Loir River and its tributaries. Six of them are in the immediate vicinity of the Loir Valley while the other four are only a few kilometres from it around its tributaries. Those clues of workshops and flint stone extraction sites localized through surface gathering are characterized by a large amount of knapping wastes but also by the discovery of bifacial pieces associated to numerous flint stone flakes and quarrying debris.

The different sites and sites' clues are located near the Loir River on the bank on both sides of the Loir Valley (fig.2). They follow most exclusively the clay outcrops. Some sites are located along small tributaries valley of the Loir River. Those are periodically dry valley which cut clay formation whose flint stones are used to axes production. The eroding caused by rains most likely allowed the surfacing of flint stone blocks, letting the Neolithic populations found their mines their. We were not able, for the time being, to find clue hitting to workshop located outside of raw material area.

We have a really small knowledge of the techniques used for flint stones extraction in this area, we can only rely on one digging operation made and furthermore on a small site. However we discovered the occurrence of circular slight depressions of a few meters diameter and also alveoli of various sizes implemented in the *thalweg* small periodical secondary tributaries to the Loir River. We can mostly likely identify these clues as remains of extraction pit, which remained visible to this day because of the protection granted in the valley slopes by trees.

At Pezou « La Chenevière-Dieu », three different flint stones extraction structure have been identified. Those are sub-rectangular shaped pit 3 to meters long and about one and a half meter deep, the bottoms of those pits are in contact with the upper bench of Campanian Era flint stones from which nodes were extracted. The flint blocks are organised in benches, embedded in clay originating from the alteration of secondary era chalk. The extraction pits are filled with waste material from knapping and directly associated with two massive piles (Despriée and Lethrosne tbp).

## 5. Flint axes production

We have recently decided to continue the technological study the Pezou mine and its important stock of artefacts (Lethrosne 2012, Despriée and Lethrosne tbp). The stone industry found in the pit filling is in shape of a pile of knapping waste including a few thousand flakes, raw blocks of flint or parts of some, cores, bifacials, totalling to a weight of more than 800 kg. Aside from a few sandstone fragments found, a single flint material was dug in this mine, coming from flint blocks from the Campanian Era.

On this mining site aside from thirty roughouts, we were able to find more than two thousands full flakes and noted their thickness and their cortical areas (Augereau 1995). We linked them to a to single object production of the *chaîne opératoire*. Production was fully focused to getting axes' blades roughouts. Extraction was exclusively linked to axes production. The technological study of those flakes distributed according to the different steps of the operating chain (Lethrosne 2012), shows a lack of flakes from of the final step of axe regulation (fig.3). Even without any other similar sites studied, this can be explained by formulating the hypothesis that the different step of the operating chain are geographical dependant in the Loir Valley. The mining sites are where the extraction and the first step of shaping out are made whereas the regulation steps are made elsewhere in other workshops or in housing.

Following the reprisal of the study on the digging site of the Pezou mine, we decided to perform a technological study of bifacial found along the Loir Valley. Aside from pieces found at Pezou, the artefacts come from surface gathering. More than 390 pieces have been classified according to the step in the *chaîne opératoire* they belong to. The bifacial preparation belong to the first steps of production (fig. 4), then come roughouts (fig. 5 and 6) and finally axes almost ready to be polished (fig. 7).

Even though the vast majority of pieces found are actually waste and pieces considered non-compliant to the stone cutter and rejected (Pelegrin 1995), they study gives us knowledge on the *chaîne opératoire* of axe making. We were able to determine the different supports to shaping out bifacial: block fragments, rough blocks or large chunks.

The reasons of rejecting a piece are many (Augereau, 1995). Break linked to either a weakness in the stone material or during knapping, those are the most often. However because the vast majority of the pieces were gathered from the surface those break might have been caused after the abandon of those pieces.
The other reasons identified are, the use of an inadequate supports, a lot of removal hinged , deep or steeped. Such making the piece uneven for the different symmetric planes axial and/or transversal which are the abandon reason.

The study of the length over width ratio of the complete products shows that the modules, of the various operating chain steps, are not too dissimilar. The bifacial preparation have dimensions comprised between 9 to 23 cm long and 3 to11 cm wide (Fig. 8), module of the rough outs are between 8 to 22 cm long and 3 to 10 cm wide (Fig. 9) and module of the axes are comprised between 8 to 25 cm in length and 3 to 8 cm width (Fig. 10).

However if we look at the length at the different steps of the operating chain, it seems that it might exist two different productions varying with their length (fig.11). A first production mainly preponderant, with products between 10 and 16cm, and a second one of a less importance whose products start at 16 cm and range to 25 cm. This second production is easier to determine when we look at the length classes of knapped axes. Where the graph for roughouts and bifacial preparations have a peak between 12 and 16cm for fray to 22cm, the axes' graph presents two peaks one at between 12 and 16cm and the second one between 18 and 20cm.

This difference in length might reveal a difference in the use of those axe's blades. A main production for tools between 10 and 15 cm in length hint to a more practical use of those whereas the second production of lesser importance might correspond to object with a more symbolical value according to the gigantic size of some (25 cm!). Similar conclusions were given for Neolithic axe's blades made out of other materials (Pétrequin and *al.*, 2002).

## 6. Distribution patterns and flint axes circulation

Concomitantly with the technological study we began working on the geographical distribution of those pieces according to the different steps of the operating chain. The geographical area of the study encompasses the north of the Loir-et-Cher, the south of the Eure-et-Loir and the western part of the Loiret. In order to allow the width area of study available and as such be as representative as possible, we inventoried bifacial pieces from various collections private or belonging to museum, and so for a sixty kilometres radius. The absence of data in the west leads to an uneven spatial distribution must have been caused by a lack of data for this part of the area. Also, most conservation places are focused to the east of the production area and we were unable to look at collection in the west, then we will have to contact conservators located in the west in the Sarthe department to resolve the issue arisen from our lack of data.

The bifacial preparations correspond at the first production stage. They are found exclusively on some camps, before interpretated as possible raw materials quarring sites or workshops (Fig. 12).

Then, rougouths' distribution always shows a very high concentration on the same precedent camps. But we can also observe some little concentrations on others sites along the Loir Valley and its main tributaries localised in less of a twenty kilometres radius around (Fig. 13).

Lastly, blade axes are localized out of initial flint mines area. They are distributed in more of a 30 kilometres radius around. They mainly follow the Loir Valley and its tributaries (Fig. 14).

The first stages of the *chaîne opératoire* (bifacial preparation) are exclusively focus on the extraction and production camps. Roughouts are also concentrated in factories area, but we can observe than roughouts and axes circulate to others sites along Aigre and Brisse's valleys. These concentrations could convey others workshops presence outdoor raw material source and neolithic villages, where the last stage of the *chaîne operatoire* of axes' production could unfold, as well as polishing stage. We note a more large axes' diffusion.

To conclude, the spatial distribution of the products outlines the boundary of a mining complex focusing around the Loir Valley. This valley could be considered as a main trunk circulation and axes production's diffusion.

## 7. Conclusion

This work underlines a part of the regional economic neolithic organisation. In fact, flint working and axe blades production seem to be localized around the Loir valley. This part of the Loir Valley could be considered as a mining complex with a single production: flint blade axe. It circulates since the first stages of the *chaîne opératoire.* Our work will pursue with the study of distribution's polished axes.
More over, we would tackle the chronological question of these productions. The artefacts wich have been analysed were out of archaeological context. Both arrowheads discovered on Pezou « La Chenevière-Dieu » excavation (Despriée and Lethrosne tbp), associated at the axes' production, allows to date it from late Neolithic period. In the same way, we don't know about the settlement in connection with axes' production.

## References

Augereau, A., 1995. Les ateliers de fabrication de haches de la minière du « Grand-Bois-Marot » à Villemaur-sur-Vanne (Aube). In Pelegrin J., Richard A. (eds.), *Les mines de silex au Néolithique en Europe: avancées récentes, actes de la table ronde internationale de Vesoul, octobre 1991,* Paris, CTHS, 145-158.

Bostyn, F., Lanchon, Y., 1992. *Jablines Le Haut Château (Seine-et-Marne). Une minière de silex au Néolithique*, Paris, DAF.

Desprièe, J., 1986. L'atelier de la Chenevière-Dieu, commune de Pezou, Loir-et-Cher. In *Actes du Xe colloque interrégional sur le Néolithique, RAO,* supplément 1, 121-122.

Despriée, J., 1997, L'atelier d'extraction et de taille du silex de « la Chenevière-Dieu » à Pezou, Loir-et-Cher. In *Préhistoire, Histoire et Patrimoine en Loir-et-Cher n°1, Mélanges offerts à Claude Leymarios,* CDPA , 16-30.

Despriée, J. and Lethrosne, H., *tbp*. Minières à silex et ateliers de façonnage de haches dans la vallée du Loir vendômois, l'exemple de Pezou « La Chenevière-Dieu » (Loir-et-Cher) In Actes du 30e colloque interrégional sur le Néolithique, Tours 2011.

Lethrosne, H., 2012. La production spécialisée d'un outil domestique. L'exemple des productions de haches en silex de la minière Néolithique de Pezou « La Chenevière-Dieu » (Loir-et-Cher). In Productions domestiques *versus* spécialisées au Néolithique et au Chalcolithique, finalités techniques et fonctions sociales : questions d'échelles ?, table-ronde de la Société Préhistorique Française à l'université de Namur (Belgique), 25 mars 2011, *Bulletin de la Société Préhistorique Française*, 109, 311-327.

Manivit, J., Desprez, N., Despriée J., 1983. *Carte géologique de la France à 1/50 000, Selommes, 2020*, BRGM, Orléans.

Maricourt de, L., 1872. Les ateliers de l'âge de pierre dans le vendômois. In *Congrès Archéologique de France,* Vendôme, 26-38.

Pelegrin, J., 1995. Réflexion méthodologiques sur l'étude de séries lithiques en contexte d'ateliers ou de mines. In Pelegrin J., Richard A. (eds.), *Les mines de silex néolithiques en Europe*, actes de la table-ronde de Vesoul, 1991, CTHS, 159-165.

Pétrequin, P., Cassen, S., Croutsch, C., Errera M., 2002. La valorisation sociale des longues haches dans l'Europe néolithique. In Guilaine J. (ed.), *Matériaux, productions, circulations du Néolithique à l'âge du Bronze*, Paris, Errance, 67-98.

Saint-Venant, J., 1917. Inventaire raisonné des polissoirs néolithiques du Loir-et-Cher et des ateliers qui les alimentaient. *Bulletin de la Société Archéologique Scientifique et Littéraire du Vendômois*, 13-65.

Saint-Venant, J., 1918. Inventaire raisonné des polissoirs néolithiques du Loir-et-Cher et des ateliers qui les alimentaient. *Bulletin de la Société Archéologique Scientifique et Littéraire du Vendômois*, 7-51.

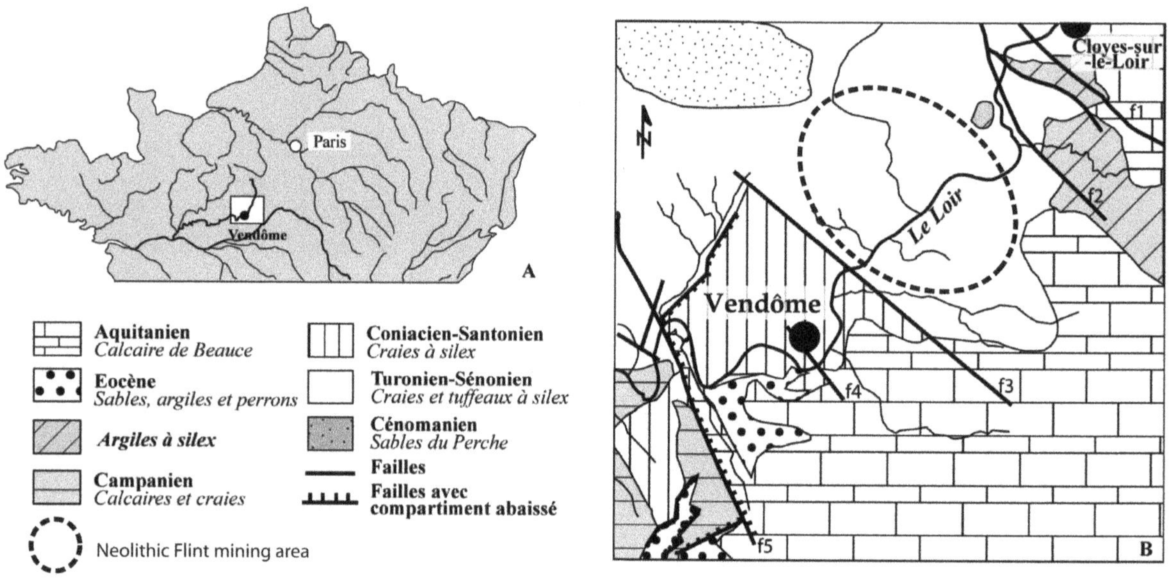


Fig. 1 : Study area. Geographical and geological context.

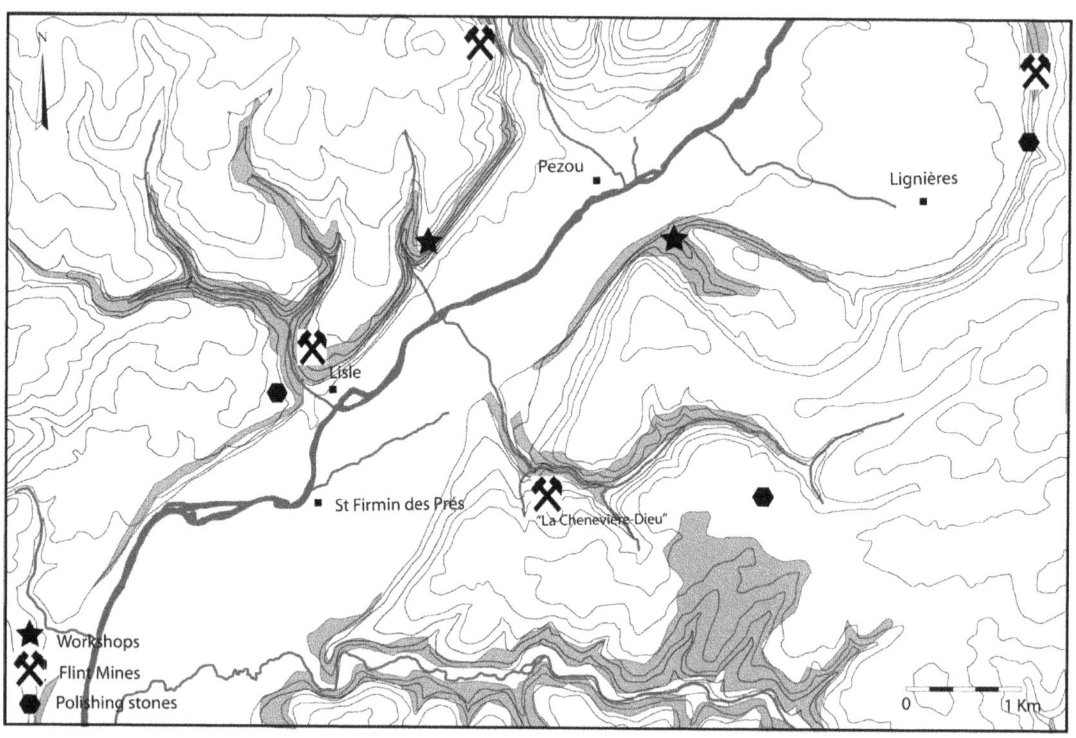


Fig. 2 : Distribution of sites and evidences for extraction and production of axes in the Loir valley. In grey, outcrops of flint clay. This raw material has been used to make axes, (geological map: Manivit and *al.*, 1983).

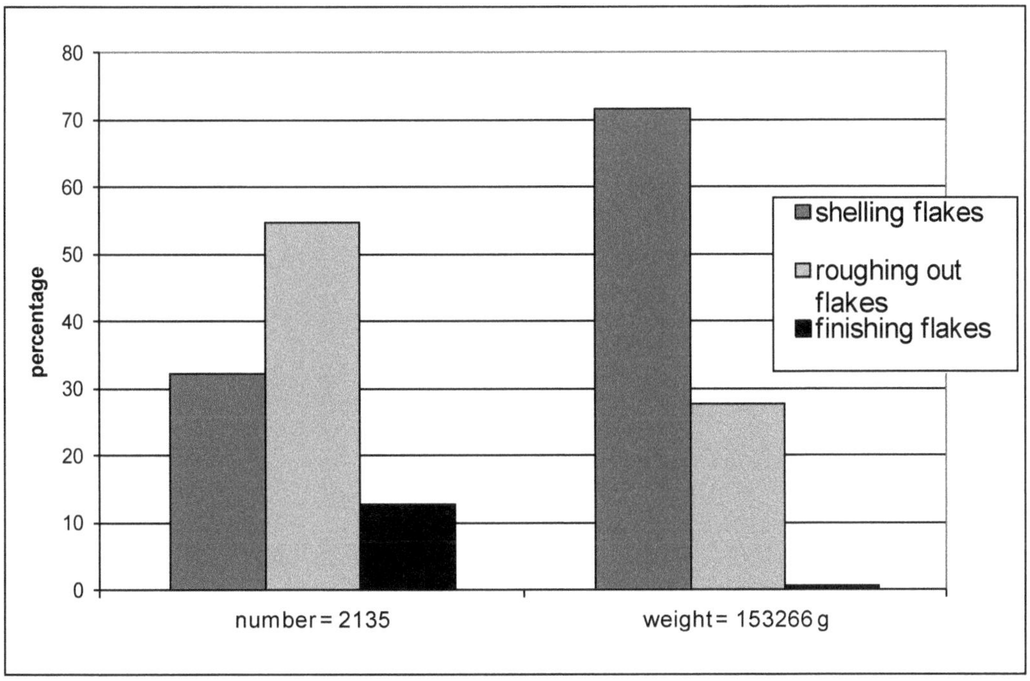


Fig. 3 : Differenciation of flakes of *chaîne operatoire* stages.

Fig. 4 : Bifacial preparations, n°1 : Pezou « Tertre Foureau », n°2 : Pezou « La Chenevière-Dieu », (coll. C. Lecubin) (photograph by J.-M. Lecoeuvre).

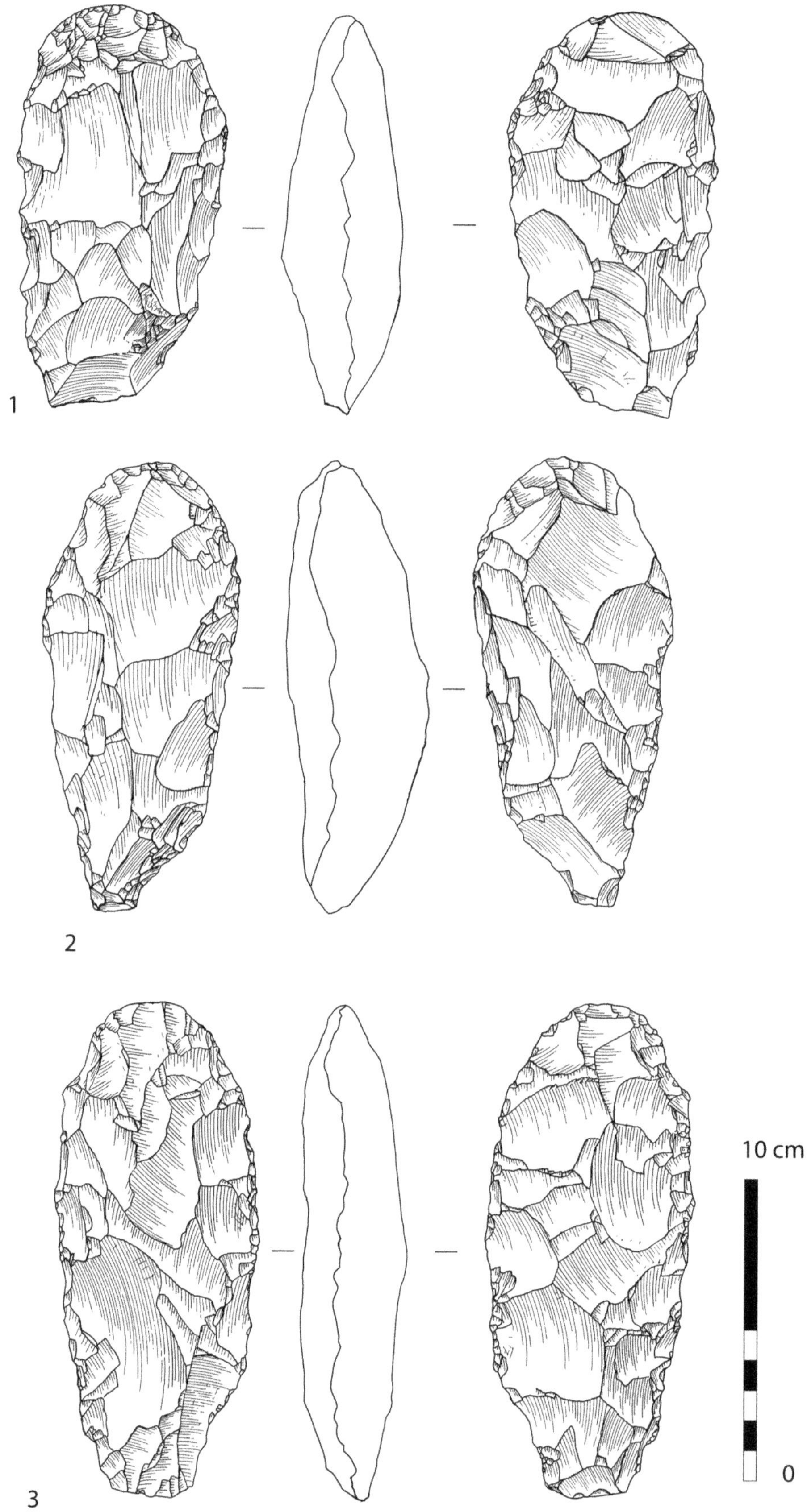


Fig. 5 : Axe blade roughouts, 1. Lisle « Viaduc », 2. Pezou « Tertre Foureau », 3. Danzé « Le Petit Boulay », (coll. C. Lecubin) (drawing by H. Lethrosne).

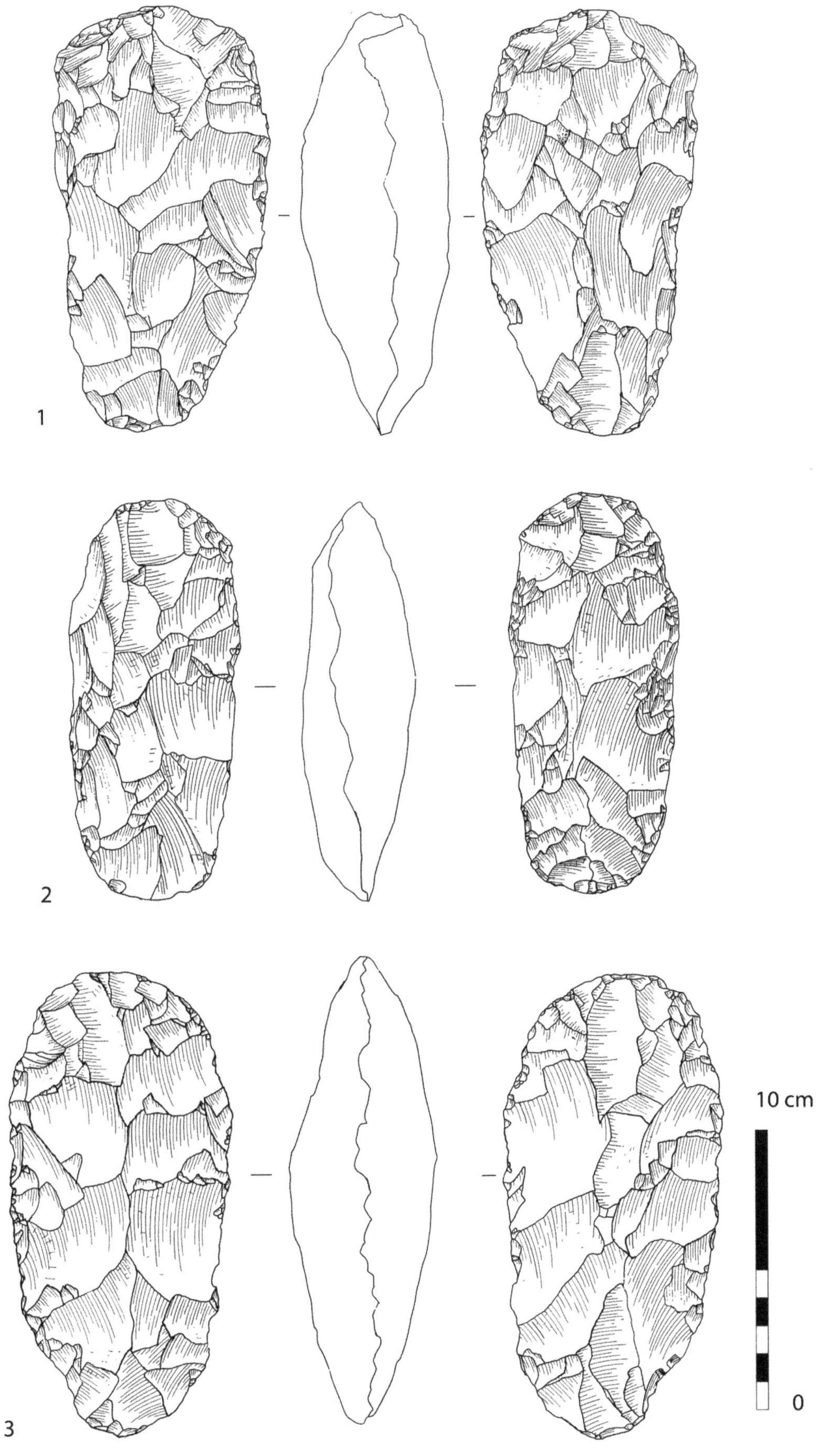


Fig. 6 : Axe blade roughouts, 1. Pezou « Tertre Foureau » (coll. C. Lecubin), 2. Pezou « Vallée de l'Aubernage » (coll. JM Lecoeuvre), 3. Pezou « Le Petit Chicheray » (coll. C. Lecubin) (drawing by H. Lethrosne).

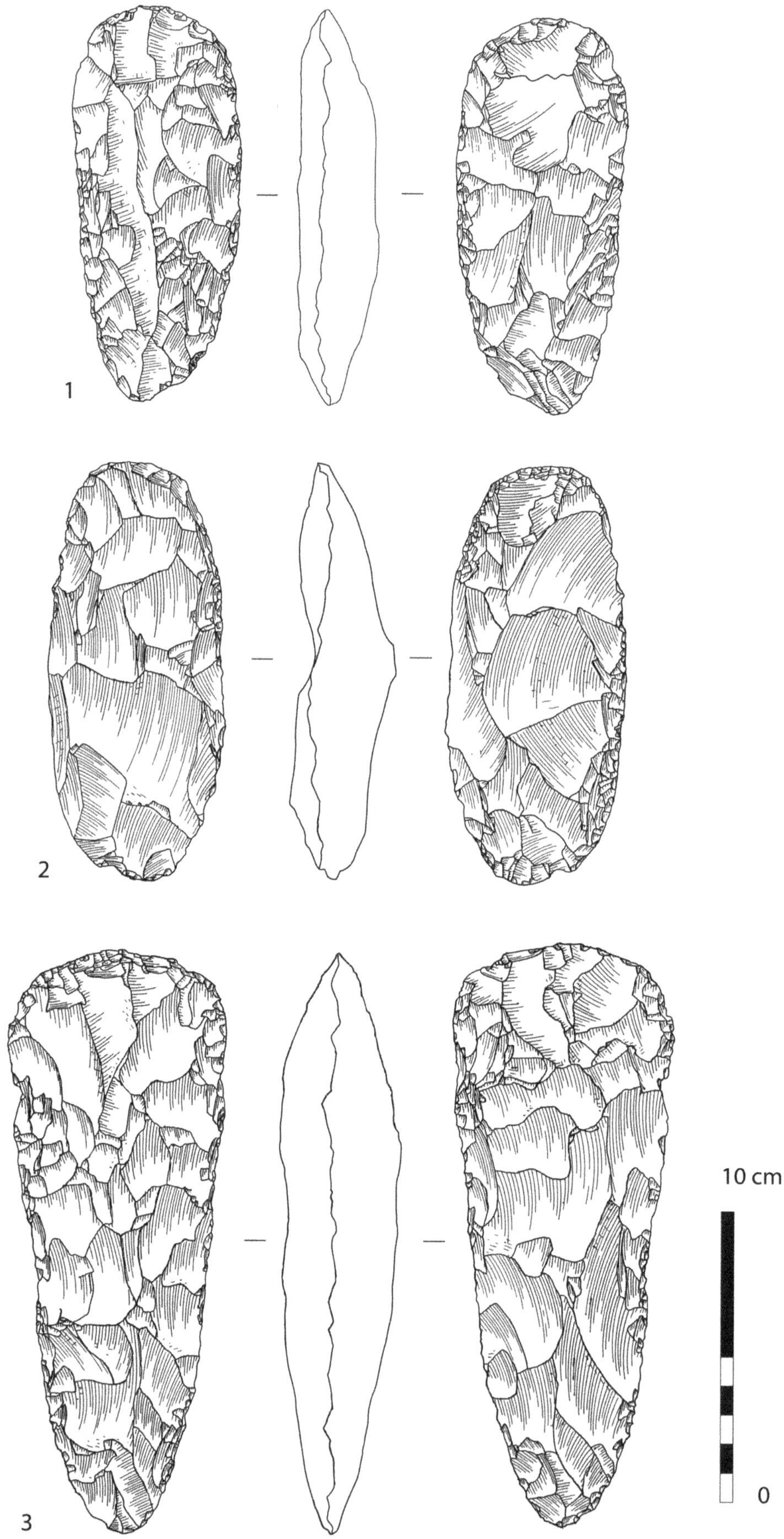


Fig. 7 : Axe blades, 1. Pezou « Chicheray », 2. Pezou « Tertre Foureau » (coll. C. Lecubin), 3. Pezou (coll. mairie de St-Firmin-des-Prés) (drawing by H. Lethrosne).

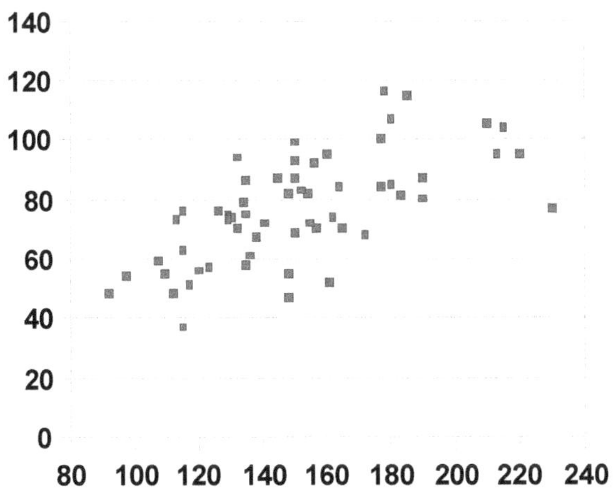


Fig. 8 : Ratio of length and weight (in mm) of bifacial preparations.

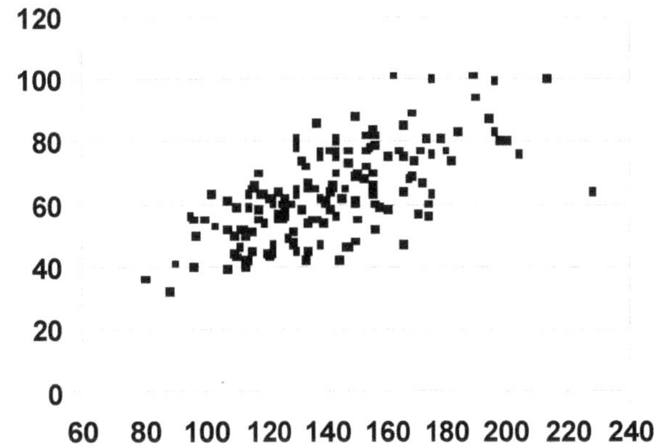


Fig. 9 : Ratio of length and weight (in mm) of axe blade roughouts.

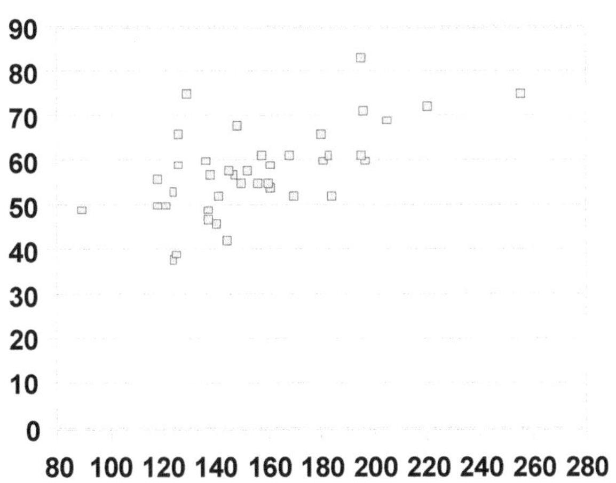


Fig. 10 : Ratio of length and weight (in mm) of axe blades.

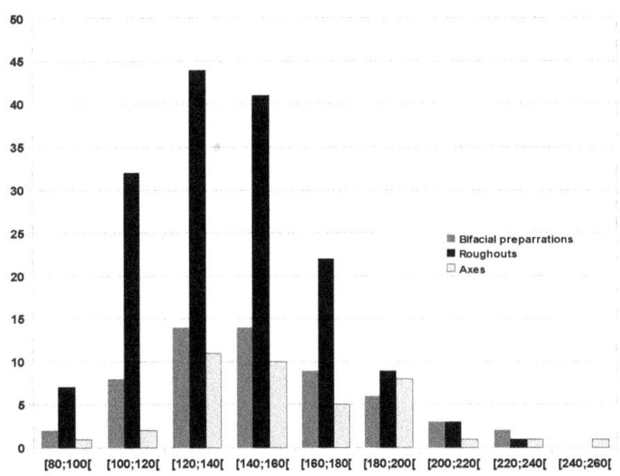


Fig. 11 : Histogramm of the length (in mm) of the products.

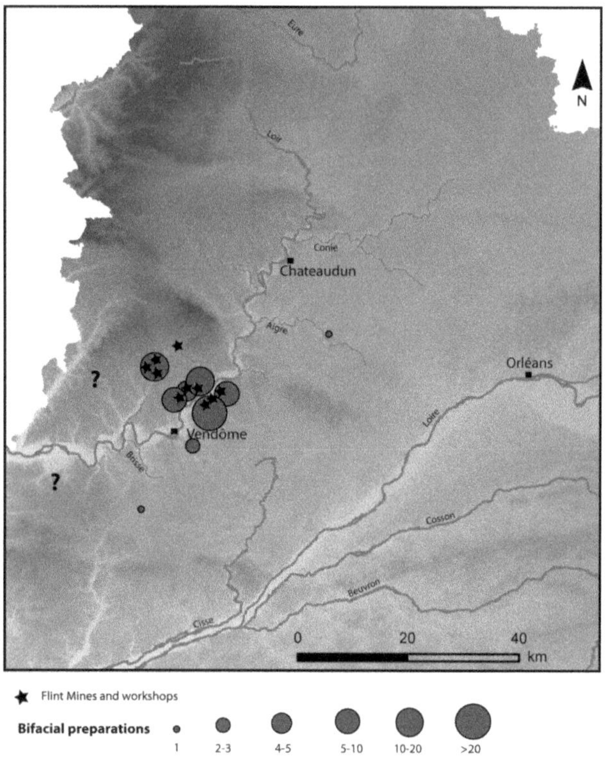


Fig. 12 : Spatial distribution of bifacial preparations.

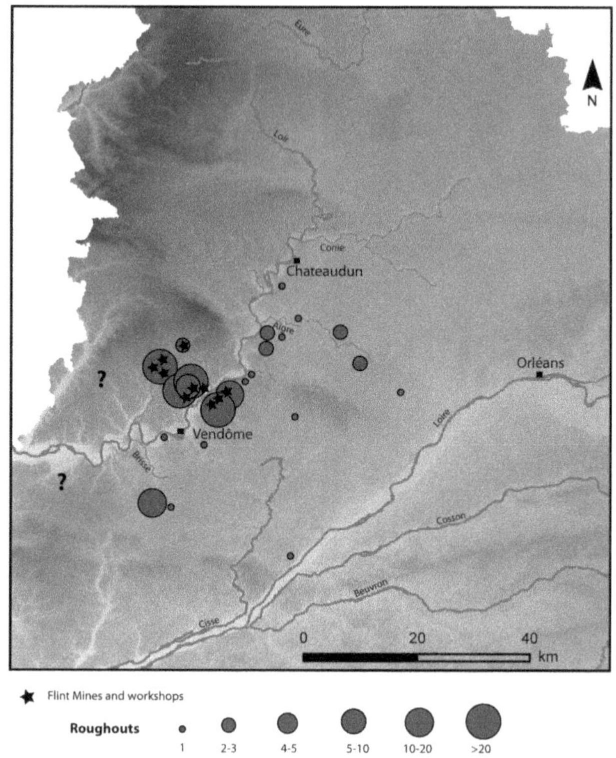


Fig. 13 : Spatial distribution of axe blade roughouts.

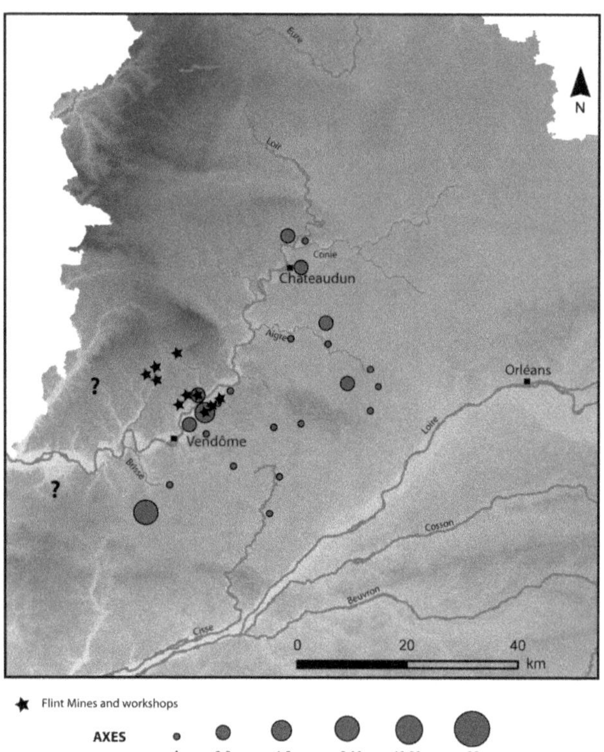


Fig. 14 : Spatial distribution of axe blades.

# The White Carpathian radiolarites – questions of chronology, extraction and distribution

Martin Oliva


## Abstract

Unlike the exploitation area of Krumlovský les (Oliva 2010), in the White Carpathians so far neither the relics of prehistoric mining nor any more substantial distribution of possible products of this extraction in the Neolithic and Eneolithic were unequivocally proved and dated. In spite of this, a major part of the abundant radiolarite industries from the Moravian side of the White Carpathians can undoubtedly be dated to the Late Neolithic or to the onset of the Eneolithic. These industries come from a special type of sites, which meet neither the criteria of a permanent agrarian occupation (lack of recessed structures, pottery and polished tools) nor the parameters of workshop localities (some km away from primary outcrops, advanced core reduction, sickle blades).

The distribution of radiolarite plays an important role only in the Old Stone Age; the evidence of presence of Palaeolithic people in the Vlára Basin, however, remains very sparse. These contradictions only contribute to an accumulation of paradoxes, which are generally associated with quarrying and distribution of lithic raw materials in exploitation areas, if solely considered as the extraction of raw materials for work tools.


**Keywords**: Carpathians; radiolarite; Middle Palaeolithic; Gravettian; Lengyel culture; quarrying; workshops; raw material distribution.

## 1. Introduction

One of the nicest and most popular raw materials for chipped tools in Central Europe was the chocolate-coloured radiolarite, the principal outcrops of which are situated in White Carpathians (Přichystal 2009, 107-110). In this sense, the region represents an exploitation area affiliated to the far more examined area of Krumlovský les. It was already Josef Skutil (1947, 31), who has drawn attention to possible evidence of mining in this part of the Carpathians; the extraction of radiolarite, however, is not yet proved with certainty (Cheben *et al.* 1995, Cheben and Cheben 2010). Surface evidence of extraction pits is more extensive and more distinctly shaped than is usual with relics of prehistoric mining, and the activity that was carried out within this area cannot yet be distinguished from the modern quarrying of stone, as proved at the examined locality of Vršatské Podhradie-Lysá (spot height 819) also by radiometric dates (Fig. 1). The amount of chipped stone industry in pits and in their neighbourhood is negligible and we even do not notice it at all during an ordinary excursion to these places. Unlike the verified exploitation zones, we can record here above all a lack of cores or tested raw material nodules and relevant cortical flakes. From an extraction pit near the spot height 819 come only 2 core-like pieces, and about 40% of industry exhibit patina marks (Cheben and Cheben 2010, 41). That does not really testify to its being associated with supposed Neolithic or Eneolithic extraction. I nevertheless assume that the discovery of indisputable evidence of prehistoric mining will not be long in coming.

A similarly disputable problem with archaeology on the Moravian side of the White Carpathians, namely the age of radiolarite industries, has already been solved to a considerable extent. The first finds by MUDr. Alois Richter were discussed in a series of short papers by Josef Skutil (1941; 1947 among others). In his pre-war articles he has referred to this mineral as to jasper, as was also done by Karel Absolon. This "jasper" (the term radiolarite was first used by K. Zapletal 1945, 220) was then known as a raw material for Palaeolithic industries, whereas in the Neolithic it was missing (and in principle it is so until today), so that until the 1950s nobody was in doubt about the new phenomenon of the "Vlára Palaeolithic". Among foreign researchers it was even Lothar Zotz (1951, 227) who has accepted the belonging of this industry to Aurignacian. A turning point in this regard represented the reaction of S. Vencl (1967) to the PhD thesis by J. Skutil (1963) based on the classification of almost 17 thousand artefacts, stored mostly in Valašské Klobouky: the loess in Sidonie would be evidently relocated on the slope, burins would have sporadically occurred as well in the Neolithic, the absence of pottery and polished industry would be typical of workshops, of course except geometrical sickle blades which cannot be of

Mesolithic age, as supposed by Skutil (1963, 118), but rather of late Lengyel origin (cf. Vencl 1971, 79; Oliva 1990, 25-26).

## 2. Neolithic Workshops

Re-dating the Vlára Palaeolithic into Neolithic was certainly justified and since then nobody has cast doubts on it, but at the same time, it also caused an outflow of interest in a phenomenon, which is for many reasons in a way exceptional. In neither of the Moravian Late Neolithic and Early Eneolithic industries, within which these assemblages probably fall, appears such a masterly handled manufacture of bladelets, as is common with all major sites in the Vlára basin. Even though these sites have yielded thousands of pieces of debitage and hundreds of cores, all of them still prospective, the whole of Eastern Moravia is distinctly dominated by northern flints, both erratic and those from the Polish Jura. The ratio of these more distant flints is highest at the time of the Linear Pottery culture and at the turn between the early and late Moravian Painted Ware (i.e. Lengyel) culture (Janák & Přichystal 2007). In the later phase of the second-mentioned culture, when Jurassic flints were quarried in Lesser Poland (Lech 1981), their incidence in Moravia has rapidly decreased without being replaced by radiolarites, the exploitation of which (whether opencast or underground) probably falls within this period, too. This is indicated not only by an increased interest in exploitation of lithic raw materials at that time but also by the typology of sickle blades and the finds of Late Neolithic and Eneolithic polished tools, which were dispersed out of the main occupation territory of the above-mentioned cultures (Langová 1995, 105, 107).

Outside the area of the White Carpathians and the Vizovice Upland we do not know of any Neolithic or Eneolithic settlement, in which radiolarite would be predominant or at least more frequent. This raw material can prevail only on the sites very close to the Vlára river (Rudimov 7 km: Lech 1983, 17-18). Radiolarite was detected at none of the 25 Late Neolithic sites, the raw material spectra of which are quantified by P. Šída (2006, tab. 4). The situation in Slovakia is less important, since there radiolarite may originate from the gravels of the Váh river. However even there we observe its marked decrease between the LBK and the Lengyel culture (Kaczanowska 1985, 151, 176-177). Thus, we meet here with the same phenomenon as in Krumlovský les and at many other exploitation centres, where raw material has been quarried and worked on a mass scale, but the products remained on the spot. Even though the settlements along the river Vlára are abundant in sickle blades (so that they were not specialised workshops but farming villages), the incidence rate of artefacts made of other raw materials is usually much lower than one per cent (unlike the settlements next to chert deposits near Krumlovský les, where imports regularly appear: Mateiciucová 1992; 2008; Oliva 2001a). The settlement agglomeration near Krumlovský les is thereby situated much closer to its main raw material outcrops than the sites with radiolarite industry in the Vlára Pass, where the abundant deposits can only be reached after climbing some kilometres uphill.

Another conspicuous phenomenon is the concentration of all industry-rich localities with exclusive use of radiolarite and with absence of polished industry, in the valley of a single river, namely the Vlára. The highest-yielding localities are those, which are situated closest to the river, on the first terrace above. Considering the volume of museum collections and written records it seems that the yield rate of individual localities increases down the river Vlára (Štítná nad Vláří, Bylnice I and II, Svatý Štěpán, Sidonie), i.e. from the west to the east, to the inside of the Carpathians (since the river dredges into the pass against the massif). These settlements are herewith more distant from the main settlement ecumenes in the lowlands, but closer to primary sources of radiolarite on the Slovak side of the White Carpathians. These sites have never yielded any Neolithic pottery, and we also dispose of no reports that any ploughed-up prehistoric pit would be examined (unlike the settlements near Krumlovský les again) – e. g. the protocols from numerous field walks carried out by Jan Pavelčík (Uherský Brod I-IV districts, Archive of the Institute of Archaeology Brno A 22-25) do not report on any pottery fragments or polished industry from this area.

## 3. Distribution of Radiolarite during the Palaeolithic

Whereas in the Neolithic we lack any more significant evidence of the export of radiolarite, in the Palaeolithic it is the other way round. This raw material is predominant at many sites: in Eastern Moravia Uherské Hradiště-Sady and Předmostí-"workshop near the cemetery", in the Brno Region

Tvarožná and Brno-Jundrov, below the Pavlov Hills Milovice I/G and Pavlov I-NW part, and the same raw material also relatively often occurs at many other sites.

Despite the generally not very high-developed transfer of raw materials during the Middle Palaeolithic, we can record already at that time a considerable use of radiolarite, which is not comparable to any other lithic raw material. The most outstanding products are the large flat bifaces from Karolín in the Kroměříž District (Oliva 1981). From Předmostí I we know of a small biface and some side-scrapers (Oliva 2005, 16), several small Taubachian artefacts come from Předmostí II (Moncel – Svoboda 1998), from the Kůlna Cave, layer 11 (Neruda 2001, tab. 1: 4 tools, 16 waste, total 0.2%) and from the Micoquian at the same locality (Valoch 1988, obr. 18: 6; Neruda 2005), points, side-scrapers and end-scrapers of the last-mentioned group are known from Bořitov V (Oliva 1987a; 2000, Fig. 7: n°1-4), and a Moustérian point was found in the Čertova díra Cave near Štramberk (Valoch 1965, 22). In Moravia, however, we do not know of any Middle Palaeolithic site, in which radiolarite would be the dominant raw material, which also partly applies to the Early Upper Palaeolithic Szeletian and Bohunician. It is, however, predominant in several Szeletian assemblages from the Váh basin in Western Slovakia (Bárta 1965). In the Bohunician, which is linked to the Stránská skála cherts, radiolarite only occurs as a complementary raw material; e.g. the stratified assemblage Stránská skála III yielded 70 artefacts comprising 18 tools and no core (Svoboda 1987, 33). In the Moravian Szeletian, very varied as far as petrography is concerned, we would hardly find any large assemblage from Moravia, in which radiolarite would be absent. However, it never counts among the three most frequent raw materials comprising the Krumlovský les chert, spongolite and northern flints. Radiolarite was widely used above all in the manufacture of leaf-points, but even among them, it is never predominant. It was mainly in the form of these points, which were undoubtedly often distributed by handing them over, that it has reached other exploitation areas, e.g. the neighbourhood of Krumlovský les. From this raw material the most outstanding triangular point (Fig. 6, n°2) within the frame of the Eastern Moravian Míškovice type was made, approximate to the forms of the Eastern European Streletskaya-Sungir culture and at the same time also to the points from Moravany-Dlhá in the Váh basin, which were probably entirely made of radiolarite (Bárta 1960; Nemergut 2010).

The incidence of radiolarite in Aurignacian is very varied. Both localities at which this raw material is predominant are situated in the Brno Region, at a distance of about 100 km from sources: developed Aurignacian from Tvarožná (more than 80%, Oliva 1987b) and possible Epi-Aurignacian from Brno-Jundrov (Oliva 1987b, 26; 1991). The incidence rate of radiolarite in this area is otherwise on average lower than in Eastern Moravia where, closer to the deposits, it is never predominant but at many localities represents the second most frequent raw material after erratic flint (Karolín I, Kvasice I and II, Nová Dědina I and II). At other rich sites, however, it is entirely absent (Žlutava VIII-Dubová). Radiolarite cannot be found at the locality of Lhotka with extremely nosed end-scrapers where erratic flint is accompanied by a relatively high amount of Jurassic chocolate flint from Central Poland (Oliva 2002, 567).

The use of radiolarite in the Moravian Gravettian is rather balanced. We can record this raw material in almost every large assemblage, but none of those collections, in which it is dominant, come from Eastern Moravia, i.e. from the area situated closer to the deposits. Radiolarite represents by far the most frequent raw material (55% excluding the radiolarite chips) in Milovice I/G (Oliva *et al.* 2009) and is predominant in the NW part of the agglomeration Pavlov I (Verpoorte 1997). It is also frequent at Dolní Věstonice I where three depositions of selected radiolarite artefacts, assorted by colour varieties, were found hidden below mammoth bones (Absolon 1938, 19, 31-32, 66-67; Oliva 2007, 18), and a workshop for radiolarite processing was discovered below an accumulation of mammoth bones (Klíma 1969; Oliva 2007, 29-31). A much larger radiolarite workshop with stone anvils was discovered in 1927 near the NE corner of a cemetery within the settlement agglomeration Předmostí (on history Absolon – Klíma 1977, 57; on industry Oliva 2007, 95-96). The most massive influx of radiolarite was of course directed at these major settlement agglomerations, even though – with regard to a much larger total amount of all artefacts – it did not become generally predominant. In accordance with this, the radiolarite in Předmostí, Dolní Věstonice I and Pavlov I has been processed in workshops from the initial phases of core preparation and reduction, and these workshops are clearly delimited in space within the settlement area. In Milovice I, on the other hand, where generally predominant, radiolarite was treated more economically, as a distant raw material. The crucial Gravettian site in the Morava basin was probably the settlement agglomeration Napajedla I on a promontory above the river, which is unfortunately known only from surface finds (Oliva 2007). To this place too, radiolarite was brought in the form of raw nodules, as proved by the large amount of debitage, unused blades, and by a higher proportion of this raw material with cores (40%) than with tools (26%). Workshops were mostly situated on the eastern slope of the promontory falling down to the river Morava. An entirely opposing situation was recorded at Boršice (old collection, Oliva 2007) – from radiolarite were made 5.6% of cores and 10% of tools. At Jarošov-"Podvršťa" – the most important site in the Morava basin, which was examined using modern methods – radiolarite is entirely outshone by the Jurassic Krakow flint (Škrdla 2005), even though the locality is situated on the eastern bank of the river, which is the closer one to the outcrops. On the Slovak side, the Vlára Pass opens to the surroundings of Nemšová where, in contrast to the Moravian side, large Gravettian workshops for radiolarite processing were discovered, almost entirely without any other raw materials and formal tools (Bárta 1961; Cheben – Kaminská 2002).

The Carpathian raw material under review is only sporadically represented in the Epi-Aurignacian and Epi-Gravettian. In the largest settlement agglomerations near Určice and Slatinice in Central Moravia, radiolarite is most frequent with side-scrapers, which, however, may be of an earlier origin (Oliva 1984, tab. III).
The last big Upper Palaeolithic civilisation, the Magdalenian, used radiolarite in an entirely common way, and the proportion of this raw material at individual sites is quite balanced, even though never dominant. Most frequent it is in front of the Ochozská Cave (Tab. 1). At the localities in the Moravian Karst it is thereby more frequent than at sites, which are situated closer to the deposits (Hranice 0.4%, Přerov 2.2%, i.e. 1 out of 45 pieces: Škrdla – Schenk – Zapletal 2008, 255).

The situation is different in the Late Palaeolithic, when radiolarite was most used at a "Tišnovian" site on the hill Špitálky (Padělky) in Uherské Hradiště-Sady, which, compared to the other contemporary sites, is situated closest to the outcrops (ca 45 km). The assemblage of 175 artefacts comprised besides the non-patinated radiolarite also some pieces of patinated as well as non-patinated flint and common chert (Valoch 1974, 113). Three out of 19 artefacts of the Federmesser-group from the Čertova díra Cave near Štramberk were made of radiolarite (Valoch 1957, 8). In the Epi-Magdalenian from the Kůlna Cave radiolarite represents 7.8% inside the layer 4 and 7.2% inside the overlying layer 3 (Valoch 1988, 22-23). At the largest Late Palaeolithic open-air site at Dřínová near Tišnov-Předklášteří, even more distant from the deposits, radiolarite only represents 14 out of 332 artefacts, i.e. 4.2% (Klíma 1963; a much larger assemblage from the excavations by O. Kos is not included).

Thus, the pattern of radiolarite distribution, which we would regard as natural, i.e. a regular decrease in amount with distance, has worked only as late as at the very end of the Old Stone Age.

## 4. Palaeolithic finds from the area of Radiolarite deposits and their neighbourhood

The most ancient artefacts (made exclusively of radiolarite) in the Vlára basin are linked with the Middle Palaeolithic. The first smoothed core was found in the valley of the creek Brumovka, in the cellar of house No. 1052 dug out in a steep slope falling down to the watercourse, in the height of about 325 m ASL. The artefact (Fig. 2, n°1) rested between weathering deposits and the subjacent fluvial sands with small-grained gravel.
It is probably a unipolar recurrent Levalloisian core for the detachment of blade flakes.
The finding place of leaf-shaped biface (Fig. 3, n°1), broken into two parts, is referred to as Bylnice-"cihelna" (brickyard). It was situated on a very moderate SE slope N of the road to Štítná nad Vláří, immediately to the west of a small watercourse in the height of 325-330 m ASL. Stratigraphy was probably never recorded.

The brickyard in Bojkovice was situated on a moderate NW slope above the left (S) bank of the river Olšava and E of its tributary Koménka in the height of 290 m ASL. In or above the brickyard were found the followings smoothed tools: a side-scraper modified bifacially by invasive retouch, the distal right side was maybe secondarily damaged by strokes (Fig. 3: 2), a side-scraper with S-like retouched right edge on a flat flake, extensive ventrobasal fracture with rough surface (Fig. 3, n°3), a bifacial knife, probably with invasive retouch worked from both the edge and the dorsal surface. Another pointed side-scraper with extensive right-side and marginal left-side retouch was found somewhere in the neighbourhood of Slavičín. The exact finding place of some other tools of Middle Palaeolithic type is also unknown (strongly retouched nosed side-scraper with bec-shaped distal part, elongated straight side-scraper on a cortical flake). The first-mentioned unlabelled artefact was stored in a box together with other items, all designated with abbreviation Št, which means Štítná nad Vláří.

The following artefacts belong to the Upper Palaeolithic, more precisely to the Gravettian. Symmetrical backed point with cortical remnant in the left bottom was found, together with naturally pointed blade, crested blade from a conical core, and some other blades and flakes, all of patinated erratic flint. The above-mentioned artefacts were selected from a collection of maybe thousand radiolarite artefacts of the Neolithic age.

Another Gravettian site is situated in the neighbourhood of a sandpit 1 km to the north of the village Vlachovice on the W slope of the hill Ďulův kopec (spot height 395), about 15-20 m above the river Vlára in the height of ca. 350 m ASL. In 1953 a pit with flat bottom was examined, sized 144x106 cm and filled with ashy grey fill, which yielded 74 tiny artefacts (smaller or equal to 1 cm), among which 32 were made of radiolarite, 33 of flint and 9 of quartzite (Pavelčík 1961, 16). The inventory from field walks comprises 41 artefacts made of radiolarite, 17 of erratic flint and 9 of quartzite. Similar was also the composition of inventories from other field walks (Pavelčík 1962). This is the so far highest ratio of non-radiolarite raw material in the whole of the Vlára basin, which, however, does not rule out that a substantial part of radiolarite artefacts may belong to some Holocene culture. J. Skutil (1963, 140) namely also reports sickle blades from this area. The industry is described in detail by Jiří Pavelčík (1961): several burins and end-scrapers without any culturally significant elements, notches, retouched blades and a point with bifacial marginal retouch on the tip, in addition also minute blade and flake debitage. Josef Skutil (1963, 75) also reports patinated flints from (Svatá) Sidonie ("Bartlovy paseky" and "Kopanice nad nádražím"), Svatý Štěpán-"Nadříčí", Nedašov and Smolina.

The most convincing examples of artefacts in the Vlára basin are surprisingly those from the Middle Palaeolithic. This is probably associated with the relatively non-specific settlement pattern of that time. The Neanderthals have colonised both caves and open landscape, nevertheless, uplands more than lowlands. The river network did not yet play any major role with regard to still relatively rare intergroup contacts, but in uplands and highlands, on the other hand, important sources of lithic raw materials may have been found, which have made the settlement activities of these people easier readable by archaeological means and maybe also more frequent and/or permanent. The fact that radiolarite also occurs outside the area of its resources in the Carpathians, whereas other raw materials do not occur in the area of radiolarite deposits, is probably given by the weakly-developed raw material transport at that time at all (in general Binford 1973; in Moravia Oliva 2002, 556) and also by the Neanderthals' preference for unusual stones and curiosities of all kinds (Oliva 2009, 16-17). As far as the cultural identification is concerned, the Carpathian tools of Middle Palaeolithic type can be assigned to some of the groups with the Levallois technique (Brumov) and to Micoquian with bifacial tools (Bylnice, Bojkovice).

Later it was undoubtedly in Gravettian that radiolarite has been carried away from the area of its deposits with highest intensity, which is proved not only by regular incidence of this raw material over the whole of Moravia but also by the existence of specialised workshops within settlements (Předmostí, Dolní Věstonice I, Pavlov I). At that time we can rightly assume the most frequent visits to outcrops, but not necessarily the workshops as such. In the above-mentioned workshops in Gravettian settlements namely occur all phases of core reduction, so that the raw material under review has normally been brought as well in unworked form (the same way as for example the Jurassic Krakow flint was brought as far as to Southern Moravia, leaving out the Předmostí region in the Central Moravia, Oliva 2001b). Evidence of temporary Gravettian occupation in the Moravian Carpathians can be proved by several artefacts from Bylnice II (Fig. 4) and perhaps also by a not very distinctive Upper Palaeolithic industry dominated by burins (made, however, also from radiolarite) from Vlachovice.

The identification of evidence of Magdalenian hunters, by whom radiolarite has been widely used, but not processed in any specialised workshops, is being complicated by the inconspicuousness of stone tools typical of this culture – for example, if some small Magdalenian (?) assemblage, such as the one from Přerov (Škrdla et al. 2008), would be made of radiolarite and situated in the area of primary sources, nobody would consider it as an evidence of this particular civilisation.

More complicated, however, would be the explanation of the absence of the characteristic tools of both main cultures falling into the early phase of the Upper Palaeolithic, namely Szeletian and Aurignacian. In these cultures, too, separate workshops for radiolarite processing are missing, but their diagnostic tool types – leaf-points and carinated end-scrapers or busked burins – are very well recognisable in the morphological sense and are commonly made of radiolarite at many sites over the whole of Moravia. Evidence of the above-mentioned cultures in the area of primary sources could thus be identified also among radiolarite industry, not only among white-patinated (flint or chert) tools. Unusual is above all the absence of leaf-points because at least a part of them were undoubtedly weapons (see their predominance in the caves such as Szeleta, Dzeravá skala or Mamutowa, which have probably served as hunting shelters). It is for this reason that these characteristic points also often occur as isolated finds in areas, which were not continuously occupied. The discovery of isolated leaf-points in the area of the White Carpathians, in my opinion, must only be a question of time.

## 5. Conclusion

Any more detailed knowledge on Palaeolithic occupation is hindered by a lack of non-contaminated inventories from that time. Patinated artefacts have always occurred as surface finds in the context of a far more abundant radiolarite component with sickle blades, so that any detailed classification of these assemblages lacks purpose. Taking into consideration only patinated specimens and artefacts of archaic form with smooth edges, we will find out that Palaeolithic evidence is rather dispersed in the landscape, whereas the rich Neolithic and Eneolithic sites are strictly concentrated next to the river Vlára. This corresponds to the settlement strategy of many Palaeolithic cultures except Gravettian, the settlements of which are situated in the neighbourhood of rivers (Oliva 1998). In accordance with these detections are also the only two localities from the area under review, which might be classified as Gravettian, and are situated immediately above the river Vlára (Bylnice II and Vlachovice). The other finds of Upper Palaeolithic industry do not enable any cultural attribution.
As far as the Neolithic is concerned, we observe some paradoxical issues:
An extraordinarily advanced lamellar technology (Fig. 5) has no analogy at other Moravian sites from this period, although these are collections from the vicinity of the lithic exploitation area. After J. Pellegrin indirect percussion was probably used for the core reduction.
Even though lithic industries from the remote region around the Vlára river contain a large amount of cores and waste, which usually indicate workshop sites (also testified by the proximity of outcrops), there are also numbers of exploited cores, formal tools and sickle segments.

Although these industries are distinguished by the excellent technology of that time, the final products were not distributed. This makes the classical conception of "supplier" and "consumer" sites somewhat questionable, at least in some circumstances.
In spite of being rich in sickle segments testifying to agriculture, the occupied areas did not yield any evidence of residential features and pits, while shards and polished industry are entirely sporadic. These findings may presumably be attributed with some palaeoethnological connotation – at the turn between the Neolithic and Eneolithic, the secluded valley of the river Vlára was probably occupied by a group of people, who were engaged not only in ordinary agrarian work but above all in exploitation and masterly processing of radiolarite, most likely without any major contacts with the surrounding world. Thus, instead of an erroneous term "the Vlára Palaeolithic" we can speak about the Neolithic of the Vlára basin as about a culturally self-contained phenomenon, a kind of a subculture of lithic prospectors and knappers.

**Acknowledgment**
This research was supported by Grant VO 00009486202.

## References

Absolon, K. 1938. *Die Erforschung der diluvialen Mammutjäger-Station von Unter-Wisternitz in Mähren. Arbeitsbericht über das zweite Jahr 1925*, Brünn.

Absolon, K., Klíma, B. 1977. *Předmostí, ein Mammutjägerplatz in Mähren*. Praha: Academia.

Bárta, J. 1960. K problému listovitých hrotov typu Moravany-Dlhá. *Slovenská archeológia* 8, 295-324.

Bárta, J. 1961. K problematike paleolitu Bielych Karpát. *Slovenská archeológia* 9, 9-32.

Bárta, J. 1965. *Slovensko v staršej a strednej dobe kamennej*. Bratislava: Vydavatelstvo SAV.

Binford, L. R.1973. Interassemblage Variability: The Mousterian and the "Functional" Argument. In C. Renfrew (ed.), *The Explanation of Culture Change*, London: Duckworth., 227-254.

Cheben, I. and Cheben, M. 2010. Research on radiolarites of the White Carpathian klippen belt. *Slovenská archeológia* 58, 13-52.

Cheben, I. and Kaminská, Ľ. 2002. Výskum paleolitického náleziska v Nemšovej. *Slovenská archeológia* 50, 53-67.

Janák, V., Přichystal, A. 2007. Distribuce silicitů krakovsko-čenstochovské jury na Moravě a v Horním Slezsku v neolitu a na počátku eneolitu. *Památky archeologické* 98, 5-30.

Kaczanowska, M. 1985. *Rohstoffe, Technik und Typologie der Neolithischen Feuersteinindustrie im Nordteil des Flussgebietes der Mitteldonau*. PAN, Warszawa.

Klíma, B. 1963. Epipaleolitická kamenná industrie z Tišnova. *Anthropozoikum* 1, 1963, 127-164.

Klíma, B. 1971. *Archeologický výzkum plošiny před jeskyní Pekárnou*, Studie AÚ ČSAV v Brně II/1, Praha.

Kostrhun, P. 2005. Štípaná industrie magdalénienu z jeskyně Kůlny. *Acta Musei Moraviae - Časopis Moravského muzea, Scientiae sociales* 90, 79-128.

Langová, J. 1995. *Pravěké osídlení Zlínska. In: Vlastivěda moravská. Zlínsko*. Brno - Zlín: MVS a Muzeum JV Moravy.

Lech, J. 1983. Flint Work of the Early Farmers. Production Trends in Central European Chipping Industries from 4500-1200 BC. An outline. *Acta Archaeologica Carpathica* 22, 1982-1983, 5-63.

Mateiciucová, I. 1999. Štípaná kamenná industrie z vedrovického rondelu. In V. Podborský (ed.) *Pravěká sociokultovní architektura na Moravě,* Brno: Masarykova univerzita, 220-223.

Lech, J. 2008. Talking Stones: The Chipped Stone Industry in Lower Austria and Moravia and the Beginnings of the Neolithic in Central Europe (LBK), 5700-4900 BC. *Dissertationes archaeologicae brunenses/pragensesque* 4, Brno: Masarykova univerzita.

Moncel, M.H. and Svoboda, J. 1998. L'industrie lithique des niveaux eemiens de Předmostí II (Přerov, Rép. Tchéque). Fouilles de 1989-1992. *Préhistoire européenne* 12, 11-48.

Nemergut, A. 2010. Paleolitické osídlenie v Moravanoch nad Váhom-Dlhej. Výsledky výskumov Juraja Bártu z rokov 1963 a 1990. *Slovenská archeológia* 58, 183-206.

Neruda, P. 2001. Využití surovin v taubachienu z jeskyně Kůlny (vrstva 11). *Acta Musei Moraviae - Časopis Moravského muzea, Scientiae sociales* 86, 3-25.

Neruda, P. 2005. Technologie micoquienu v jeskyni Kůlně. *Acta Musei Moraviae - Časopis Moravského muzea, Scientiae sociales* 90, 23-78.

Neruda, P. and Kostrhun, P. 2002. Hranice-Velká Kobylanka. Mladopaleolitická stanice v Moravské bráně. *Acta Musei Moraviae - Časopis Moravského muzea, Scientiae sociales* 87, 2002, 105-156.

Neruda, P., Nerudová, Z., Čulíková, V. 2009. Loštice I – Kozí vrch. Magdalénienská stanice v horním Pomoraví. *Acta Musei Moraviae - Časopis Moravského muzea, Scientiae sociales* 94, 39-64.

Oliva, M. 1981. The Acheulian Finds from Karolín (Moravia). *Anthropologie* 19, 27-32.

Oliva, M. 1984. Technologie a použité suroviny štípané industrie moravského aurignacienu. *Archeologické rozhledy* 36, 601-628.

Oliva, M. 1987a. Vyvinutý micoquien z návrší Horky u Bořitova - první výsledky. *Acta Musei Moraviae - Časopis Moravského muzea, Scientiae sociales* 72, 21-44.

Oliva, M. 1987b. *Aurignacien na Moravě – L'Aurignacien en Moravie.* Studie Muzea Kroměřížska 87, Kroměříž: Muzeum Kroměřížska.

Oliva, M. 1990. Štípaná industrie kultury s moravskou malovanou keramikou v jihozápadní části Moravy.

*Acta Musei Moraviae - Časopis Moravského muzea, Scientiae sociales* 75, 17-37.

Oliva, M. 1991. Mladopaleolitická stanice s radiolaritovou industrií v Brně - Jundrově. *Acta Musei Moraviae - Časopis Moravského muzea, Scientiae sociales* 76, 19-29.

Oliva, M. 1998. Geografie moravského gravettienu. *Památky archeologické* 89, 39-63.

Oliva, M. 2000. Le Paléolithique moyen en Moravie: les industries lithiques et leurs matières premières. In: A. Ronen and M. Weinstein-Evron (eds.): Toward Modern Humans, Yabrudian and Micoquian, 61-76 . BAR int. ser. 850, Oxford 2000.

Oliva, M. 2001a. Sídliště lidu s moravskou malovanou keramikou v okolí Krumlovského lesa a jejich štípané industrie. In: *Konf. o otázkách neolitu a eneolitu našich zemí,* Mostkovice 1999. Pravěk Suppl. 8, 197-231.

Oliva, M. 2001b. Gravettienská sídliště u Pavlova. K otázce využívání silicitů krakovské jury. *Acta Musei Moraviae - Časopis Moravského muzea, Scientiae sociales* 86, 45-99

Oliva, M. 2002. Využívání krajiny a zdrojů kamenných surovin v mladém paleolitu českých zemí. *Archeologické rozhledy* 54, 555-581.

Oliva, M. 2005. *Civilizace moravského paleolitu a mezolitu.* Brno: MZM.

Oliva, M. 2007. Gravettien na Moravě. *Dissertationes archaeologicae brunenses/pragensesque* 1, Brno: Masarykova univerzita.

Oliva, M. 2009. Facts and deliberations on the earliest Art, In: K. Valoch & M. Lázničková Galetová (ed.) *Nejstarší umění střední Evropy*, Brno: MZM, 15-38, 69-71.

Oliva, M. 2010. Pravěké hornictví v Krumlovském lese. Vznik a vývoj industriálně-sakrální krajiny na jižní Moravě. Prehistoric mining in the « Krumlovský les » (Southern Moravia). Origin and development of an industrial-sacred landscape. *Anthropos* N.S. 24, Brno: MZM.

Oliva, M. (ed.) et al. 2009: Sídliště mamutího lidu u Milovic pod Pálavou. Otázka struktur s mamutími kostmi. *Anthropos* N.S. 19, Brno: MZM.

Pavelčík, J. 1961. Aurignacko-gravettská stanice u Vlachovic. Zpráva za 1. čtvrtletí 1961 Muzea J.A. Komenského, Uh. Brod, 16-20.

Pavelčík, J. 1962. Paleolitická industrie z „Ďulova kopce“ u Vlachovic. Zpráva za 2. pololetí 1962 *Muzea J. A. Komenského, Uherský Brod* 27-28.

Přichystal, A. 2009. *Kamenné suroviny v pravěku východní části střední Evropy.* Brno: Masarykova univerzita.

Skutil, J. 1941. Neue paläolithische Funde aus Mähren. *Quartär* 3, 161-168.

Skutil, J. 1947. Karpatské rádiolaritové vlárské paleolitikum moravské. *Historia slovaca* 5, Eisnerov sborník, 16-33.

Skutil, J. 1963. *Bělokarpatské radiolaritové paleolitikum Povláří.* Brno (rukopis kandidátské disertace).

Svoboda, J. 1987. Stránská skála. Bohunický typ v brněnské kotlině. *Studie archeol. ústavu ČSAV v Brně* XIV/1. Praha 1987.

Svoboda, J. (ed.) 2002. *Prehistorické jeskyně,* Dolnověstonické studie 7, Brno: ARÚ, 226-228.

Šída, P. 2006. Distribuční areály surovin v neolitu na území České republiky. *Archeologické rozhledy* 58, 407-426.

Škrdla, P. 2002. Magdalénská sídelní struktura v jižní části Moravského krasu. Problematika otevřených sídlišť. In J. Svoboda (ed.), *Prehistorické jeskyně,* Dolnověstonická studie, 7, Brno, 229-254.

Škrdla, P. 2005. The Upper Paleolithic on the Middle Course of the Morava River. *Dolnověstonické studie* 13. Brno.

Škrdla, P., Schenk, Z., Zapletal, R. 2008. Přerov (k.ú. Přerov-město, okr. Přerov). *Přehled výzkumů* 49, 254-257.

Valoch, K. 1957. Jeskyně Šipka a Čertova díra u Štramberku. I. Mladší paleolit. *Acta Musei Moraviae - Časopis Moravského muzea, Scientiae sociales* 42, 5-24.

Valoch, K. 1965. Jeskyně Šipka a Čertova díra u Štramberku. *Anthropos* N.S. 9. Brno: Moravské muzeum.

Valoch, K. 1974. Eine spätpaläolithische Industrie aus Sady bei Uherské Hradiště. *Sborník prací fil. fak. brněnské univ.* E 18-19, 1973-1974, 111-124.

Valoch, K. 1988. Die Erforschung der Kůlna-Höhle 1961-1976. Mit Beiträgen von J. Jelínek, W. G. Mook, R. Musil, E. Opravil, L. Seitl, L. Smolíková, H. Svobodová, Z. Weber. *Anthropos* N. S. 16, Brno: Moravské muzeum.

Valoch, K. 2002. Die Magdalénien-Fundstelle an der Ochoser-Höhle im Mährischen Karst. Ein Beitrag zur Problematik des Magdalénien in Mähren. In J. Svoboda (ed.), *Prehistorické jeskyně, Dolnověstonické studie* 7, Brno: ARÚ183-225.

Vencl, S. 1967. K otázce datování tzv. vlárského paleolitu. Sborník filozofickej fakulty Univerzity Komenského, *Musaica* XVIII (VII), 3-13 a 2 tab. Bratislava.

Vencl, S. 1971. Současný stav poznání postmesolitických štípaných industrií v Československu. In J. K. Kozłowski (ed.), *Z badań nad krzemieniarstwem neolitycznym i eneolitycznym*, 74-99. Kraków: Muzeum Archeologiczne.

Voláková, S. 2005. Štípaná industrie z jeskyně Pekárny (Mokrá) v Moravském krasu. *Acta Musei Moraviae - Časopis Moravského muzea, Scientiae sociales* 90, 129-160.

Verpoorte, A. 1997. Along the peripheries of a Radiolarite Concentration: The Lithic Industry of 1956/ABC and 1958. In J. Svoboda (ed.), *The Pavlov I - Northwest. The Upper Paleolithic burial and its settlement context*, Dolnověstonické studie 4, Brno: ARÚ, 211-226.

Zapletal, K. 1945. Mineralogische Beschaffenheit der Unter-Wisternitzer Steinartefakte. In K. Absolon, *Výzkum diluviální stanice lovců mamutů v Dolních Věstonicích na Pavlovských kopcích na Moravě. Pracovní zpráva za třetí rok 1926*, Brno, 219-221.

Zotz, L. 1952. *Altsteinzeitkunde Mitteleuropas.* Stuttgart: F. Enke Verlag.

| | a | b | c | d | e | f | g | h |
|---|---|---|---|---|---|---|---|---|
| | cores | flakes | blades | tools | waste | total pcs. | total % | after: |
| Moravian Karst: | | | | | | | | |
| Pekárna - inside | 1 | x | x | 5,5 | 2,1 | 1090 | 4,3 | Voláková 2005 |
| Pekárna - outside | x | x | x | x | x | 148 | 3,1 | Klíma 1974 |
| Kůlna Level 5 | 6,5 | 1,8 | | 6 | 1,3 | 17 | 2,2 | Kostrhun 2005 |
| Kůlna Level 6 | 0 | 1,7 | | 3,3 | 0,3 | 16 | 1,1 | Kostrhun 2005 |
| Balcarka | x | x | x | x | x | 28 | 8,7 | Nerudová - Neruda 2010 |
| Býčí skála | 20 | 5,1 | 8,2 | 8 | 5,3 | 111 | 7,9 | Oliva 1995 |
| Ochozská | 10 | 6,9 | 8,3 | 7,8 | x | 116 | 17 | Přichystal 2002; Valoch 2002 |
| Mokrá - Quarry V | 0 | 0 | 0 | 0 | 0 | 0 | 0 | Škrdla 2002 |
| N and E Moravia: | | | | | | | | |
| Hranice | 1 | 0,7 | | 0,4 | 0,2 | 12 | 0,4 | Neruda - Kostrhun 2002 |
| Přerov | 0 | 0 | 0 | 0 | 0 | 0 | 0 | Škrdla et al. 2008 |
| Loštice | 0 | 0 | 0 | 0 | 0 | 0 | 0 | Neruda et al. 2009 |

Tab. 1 : Radiolarite in the Magdalenian of Moravia.

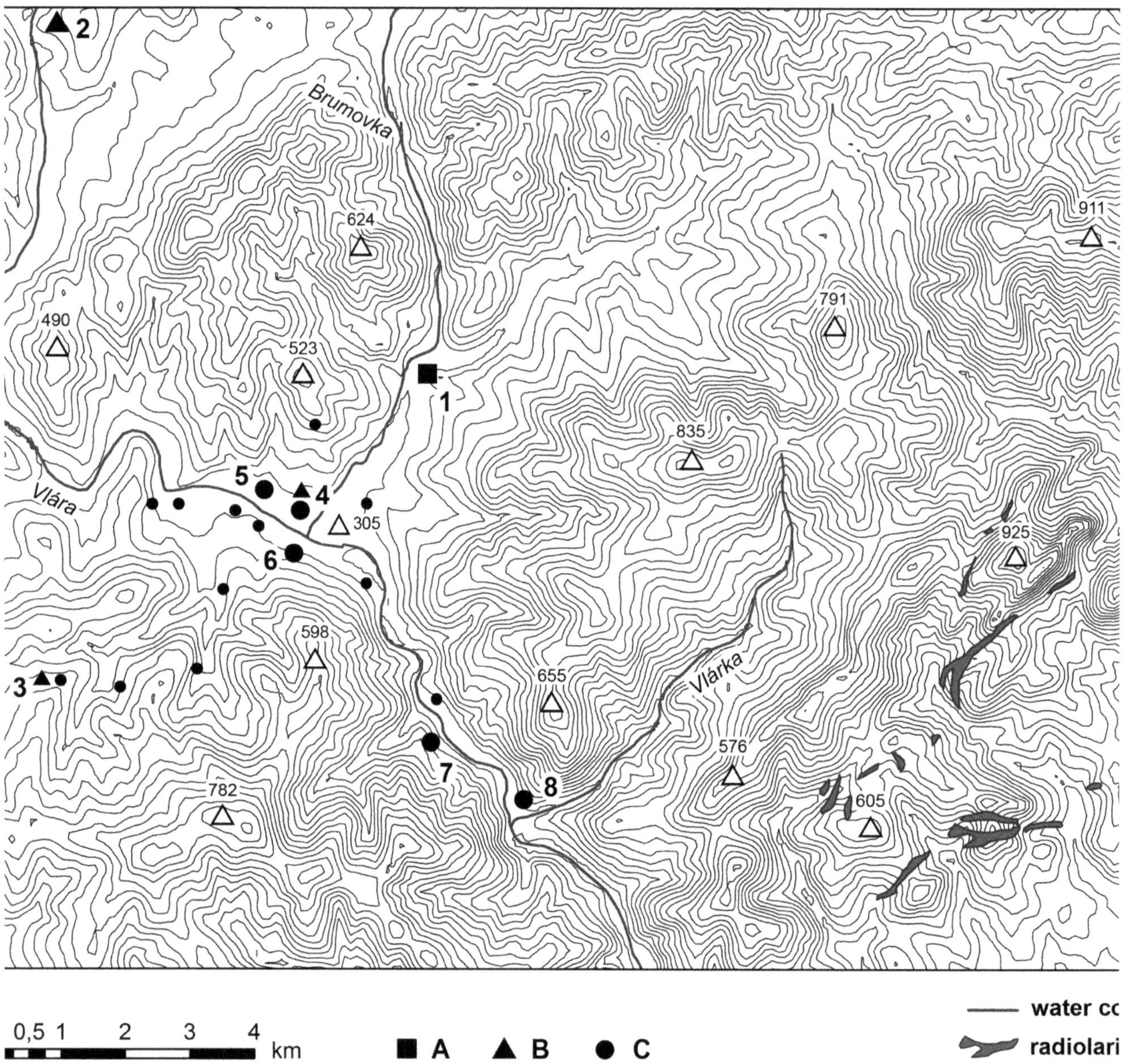


Fig. 1 : Sites with radiolarite industries in the Vlára Pass. A Middle Palaeolithic, B Gravettian – Epigravice II, 5 Bylnice I, Štítná nad Vláří, 7 Svatý Štěpán, 8 (Svatá) Sidonie. Near the point 925 Chmeľová possible quarrying of the radiolarite.

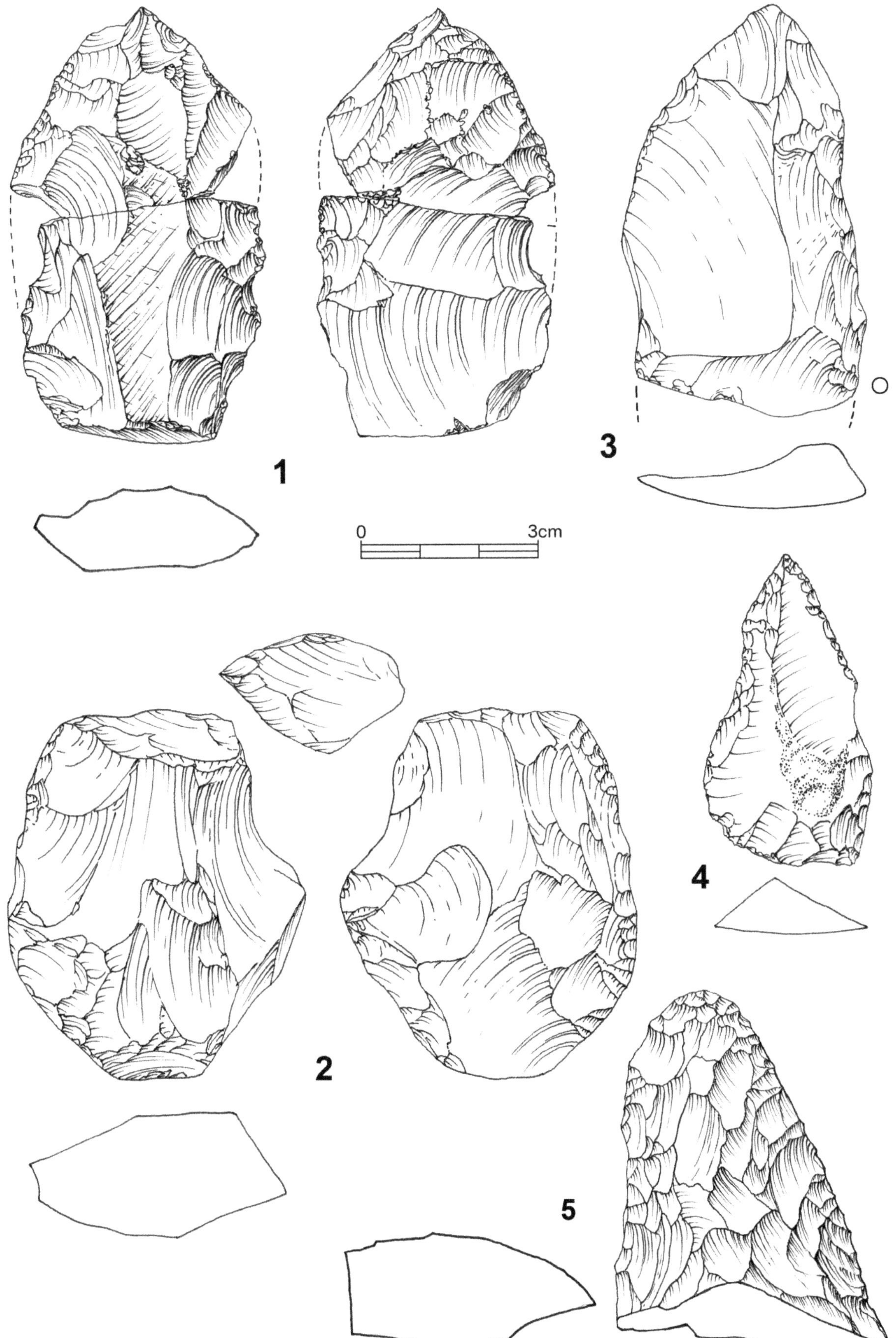


Fig. 2 : 1. Brumov-Bylnice, 2. unknown locality, Levalloisian cores of radiolarite.

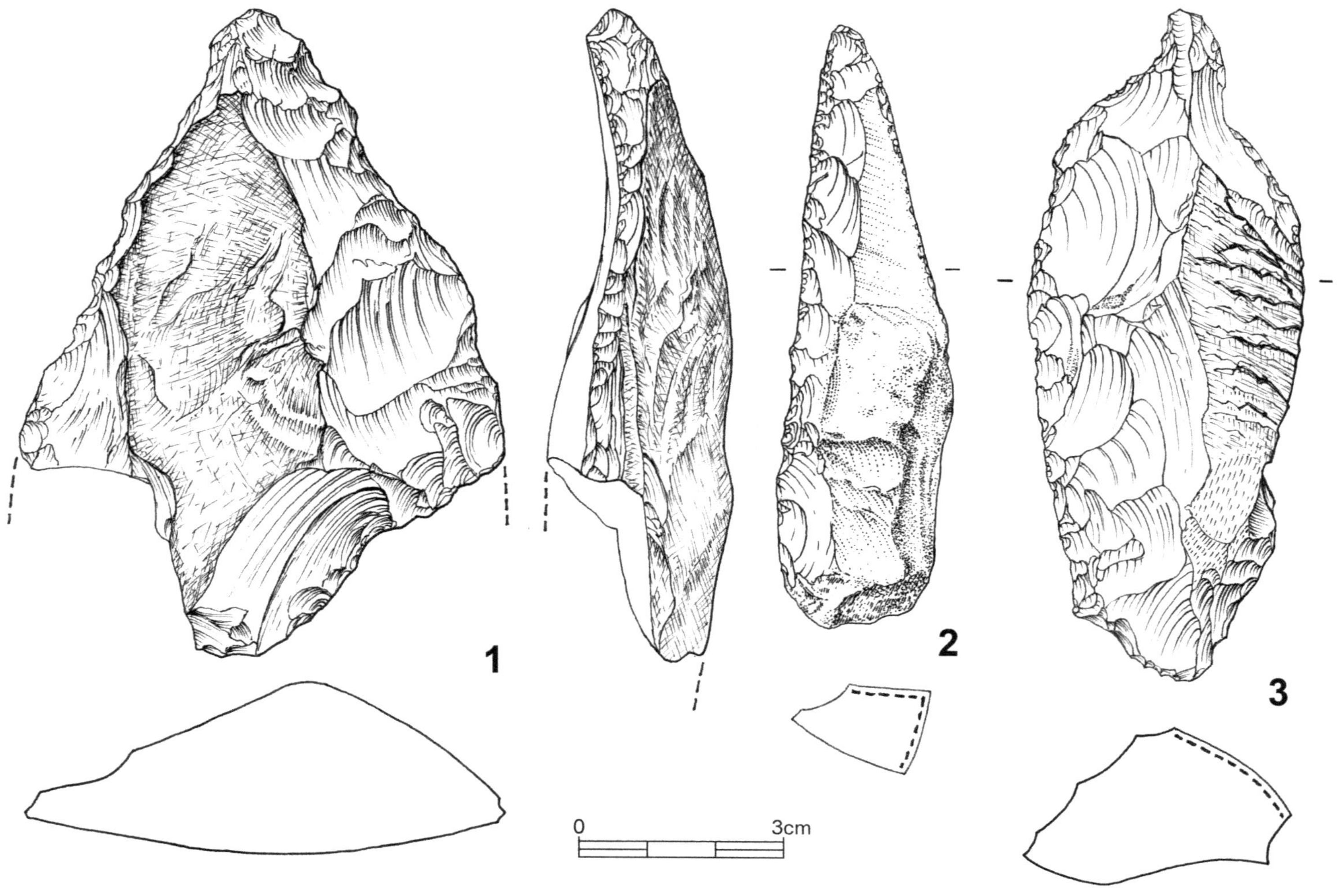


Fig. 3 : 1. Brumov-Bylnice – loam pit, 2-3. Bojkovice – „Nad cihelnou“,, Middle Palaeolithic tools of radiolarite.

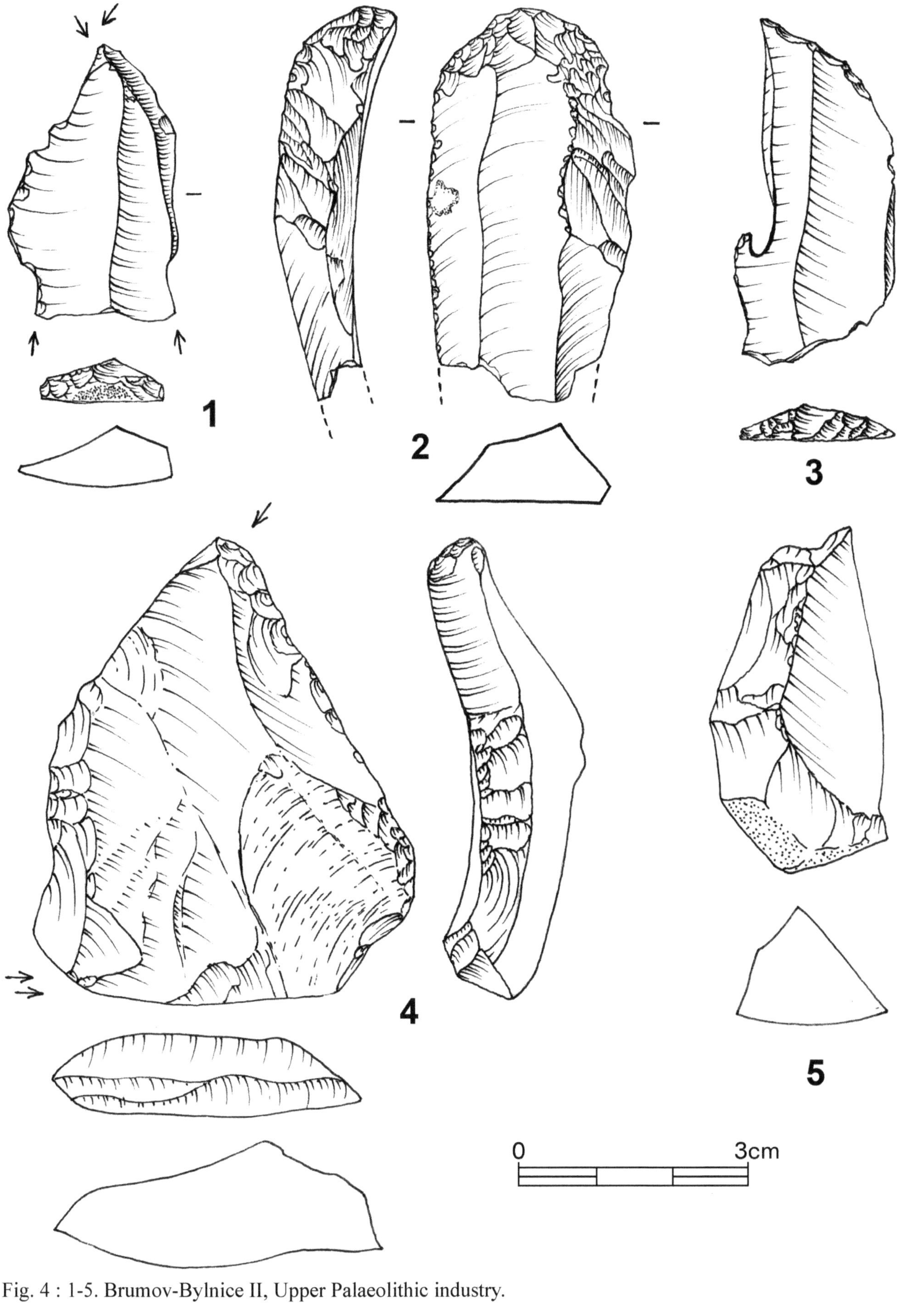


Fig. 4 : 1-5. Brumov-Bylnice II, Upper Palaeolithic industry.

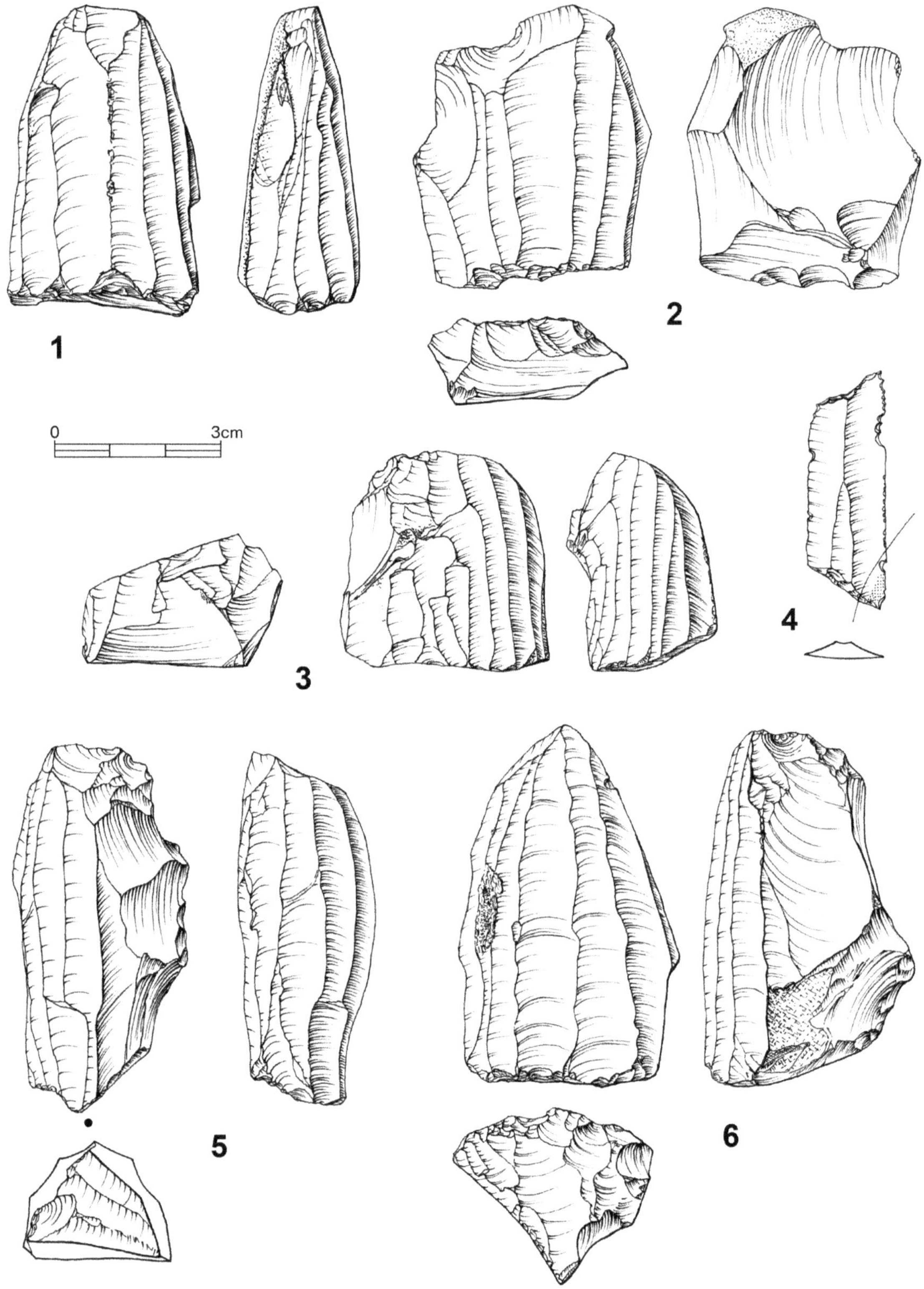


Fig. 5 : Bladelet cores and a sickle blade (4) from Neolithic sites in Vlára Pass. 1. Štítná nad Vláří – Valentovy paseky, 2 .Štítná nad Vláří – Kršlisko, 3-4. Sidonie. 5-6. Štítná nad Vláří.

# The Exploration of a Mining Site for Radiolarite in the White Carpathians Area

Ivan CHEBEN and Michal CHEBEN

## Abstract

An archaeological investigation of mining pits above Horná lysá in the area of Chmelová Hill in Slovakia gives evidence of radiolarite extraction. It probably dates to the Neolithic, unless radiocarbon datation has not still been made. Fire mining was used to process the rock layers in which the radiolarite was outcropped. In two pits, 92 stone artefacts where found, most of them are flakes and waste, but pre-cores and cores and blades are also present.

**Keywords**: mining pit; radiolarite; chipped stone industry; western Slovakia; Neolithic; Eneolithic.

Radiolarite was one of the most important sources of silicite in both the Neolithic and Eneolithic. Its primary occurrence is associated with the mountain belt of the Carpathian arc. Its sources are found in the White Carpathians, where radiolarite was presumably mined. In the course of the mapping of its primary sources, archaeologists discovered, apart from a previously researched mine in Bolešov, prehistoric mining sites in four more locations. On the hill over the site Horná lysá in Vršatské Podhradie (Ilava District, Slovakia) a source of radiolarite is found immediately below the ridge. A circular, cauldron-shaped pit was excavated, containing layers of un-mined radiolarite. Exploration of the eastern section of the pit has confirmed that it is a pingo (hydrolaccolith) in which radiolarite was mined. Further evidence is provided by the excavated artefacts; although the majority of them are production waste, there are also pre-core shapes, flakes and blades.

Based on the analyses of find complexes of chipped stone industry from the Neolithic and Eneolithic sites in Western Slovakia, it may be stated that radiolarite was one of the most frequently used domestic silicite raw materials whose primary sources can be found in the klippen belt of the Carpathian arc. It was the exploration of radiolarite which earned increased attention during the study of artefacts from the Palaeolithic Age, including the places of its primary occurrence (Bárta 1961; 1979; Mišík 1999). It is only natural that alongside these issues the researchers were gradually focusing also on the question of the search for suitable sites and evidence of mining radiolarite, or hornstone, from primary sources, as well as primary processing of the raw material in the immediate, or wider, area of its occurrence.

The beginnings of the surface or deep mining of silicites in the wider area of central Europe go back to the beginning of the Neolithic Age and are attributed to the Linear Pottery culture. Since the beginning of the second half of the twentieth century the White Carpathians area (Vencl 1967), between the Vlára River and Vršatské Podhradie (Ilava District), has been known for find assemblages of Palaeolithic, Neolithic and Eneolithic chipped stone industry made of radiolarite from the klippen belt of the White Carpathians. During several recent field explorations into the observed area the occurrence of the raw material fragments (radiolarite and hornstone) has been localised, as well as, exceptionally, several types of artefacts or production waste in more than twenty sites (Cheben *et al.* 1995). The sites′ common feature is the fact that there occur mostly splinters and flakes, but also pre-core shapes, many times with the remnant of the mother rock. In the area of the primary occurrence of radiolarite a sporadic blade can be found, or an artefact with a retouching.

Within the wider surrounding of Vršatské Podhradie a potential primary source of radiolarite was searched for also in the area of Chmelová Hill, but the expectation was not confirmed. During one of the field explorations in 2004 a new site was found in the observed area, where radiolarite was obtained from the primary source. In the hill area above Horná lysá (Fig. 1, n°1) the occurrence of radiolarite was identified in two places. On the ridge itself, where on its south-western and north-eastern edge two marked pits are situated. In the first case it is a cauldron-shaped sunk pit of a circular ground plan (Fig. 1, n°2), which is already on the slope more steeply falling towards south-east, and is situated closer to the edge of the revealed limestone outlier. The pit itself was filled by the soil mixed with fragments of limestone, including fragments of raw radiolarite mostly of brown and green colour. The second presumably mining pit is situated on the opposite end of the 200 m long ridge. It is of an irregular shape and was hollowed out also at the edge of the limestone outlier. In addition to these two pits, on the western slope, 50–60 under the ridge, a marked layer of radiolarite of mostly brown, red-brown and green colour was caught. In spite of the fact that in this part the slope is relatively steep, the configuration of the terrain permits assumption that the more distinctively observable depression with a slightly terraced lower part originated as a result of the activity during the mining of radiolarite.

During August 2008 a short-term exploratory research on the above described pingo was carried out. In this case the primary aim was to confirm assumptions that it is a prehistoric mining artefact. The pingo itself, with 550 cm diameter, had a funnel-like shape and a rounded lower part. Since the pingo was situated on the southern, relatively steep slope of the hill´s top, its total vertical distance between the current southern and northern edge is more than 250 cm. The research was concentrated on the unearthing of the eastern half of the pingo, caused by the fact that this was the part in which was preserved a remnant of un-mined silicite raw material which made up the pit´s wall. The gradual unearthing in the main pingo (labelled as II) cleared both a part of the eastern wall with the layer of radiolarite (Fig. 1, n°3), as well as a part of the outlier in the lower part of the pingo (Fig. 1, n°4). The original bedding of the rock was partially preserved in the middle of the northern edge of the pit, and it could be observed up to the excavation level. Since in the course of the exploration it was found out that pingo II is significantly deeper than it was originally assumed, its research remained, also because of safety reasons, uncompleted.

By gradual deepening of the layers at the north-eastern edge a part of, probably, an exploratory pingo (labelled as I), was captured, which at the western edge followed the vertically or obliquely, placed layers of limestone – the outlier. The recorded situation shows that radiolarite in this part was mined also with the help of fire. This is proved not only by several artefacts damaged by fire (Fig. 2, n°8), but also by the rock occasionally burnt to a red colour. Small parts of wooden charcoals were identified in the layer situated in the depth of 140 cm. Moreover, it has to be emphasised that in the 100–150 cm deep layer a substantial part of the chipped industry inventory was situated.

A different situation was identified in the southern part of pingo II, where a backfill made up of fragments of radiolarite from the peripheral parts and broken limestone (Fig. 1, n°5) was found. Our preliminary assumption is that this part of the pingo had been mined earlier and consequently filled with unneeded "waste rock" during the progression of mining works northwards.

The exploration of the eastern half of the pingo brought a relatively representative assemblage of chipped industry, mostly coming from the pingo I backfill. Based on the acquired information, one may state that it is in fact a smaller exploratory or inquiry pingo which was situated at the northern edge of main mining pingo II. In general processing of database and primary assessment of the acquired assemblage of chipped industry we drew not only on the traditional typological-morphological analysis, but also followed the signs of a more modern, so-called dynamic-technological approach to the evaluation of artefacts (Dzieduszycka-Machnikowa – Lech 1976; Oliva 1998). Within the evaluated assemblage four classification classes could be distinguished: a) pre-core shapes and cores; b) blades and their fragments; c) flakes and waste; d) instruments.

The chipped industry assemblage from the area of the eastern half of the pingo is made up by 106 artefacts (Fig. 2), which were divided, based on the evaluation criteria, into three main production categories, with artefacts shorter than 12 mm included to general analysis as well. According to these criteria, the chipped industry assemblage shows unequal representation in individual groups. Into the first category, of pre-core shapes and cores, were included two artefacts. A somewhat more numerous category, containing twelve specimens, was made up by blades and their fragments. The nature of the explored object fully confirmed that the most numerous group is made up by the category of flakes and waste. As many as 92 artefacts were included into this group. With respect to the fact that it was the exploration of a mining work intended for the extraction of radiolarite, it may be stated that the artefacts made of radiolarite from the klippen belt are represented by 99 % in the evaluated inventory. In one case a flake was obtained, made of silicite from glacigene sediments and most probably brought to the pingo area during excavation works. As sources for its exploitation served glacial moraines and terraces of rivers in northern Moravia and Silesia as well as in the southern parts of Poland where Pleistocene glaciers stretched.

It is important to note that the majority of evaluated artefacts of the chipped industry assemblage was patinated and that that some specimens, as well as several pieces of radiolarite raw material from the backfills of both pingos, showed traces of burning-up, or faced the impact of fire. In several places it was noticeable that also the mother rock (limestone) showed traces of burning-up, or the impact of fire. As regards this fact, one may rightfully assume that in radiolarite mining fire was used for the erosion of mother rock.

The category of pre-core shapes and cores is represented by two specimens in the evaluated assemblage. The backfills of both unearthed parts of pingos were made up by pieces of radiolarite of various sizes, but did not carry traces of intentional processing. Only in one case a fragment of radiolarite was found, showing traces of brush-off. Based on the preserved traces of (short-term?) use, it could be utilised as a massive side scarper. From pingo I several preparation flakes have been documented, allowing one to assume that the production of pre-core forms and cores prepared for the brushing-off of blades was done directly in the pingo or in its close surrounding. In pingo I a core was also found, which was unsuitable because of technological error during the preparation of the core base and thus it was thrown away as a defect.

In the evaluated assemblage, the category of blades and their fragments contains a relatively small number of items. In total, there are nine blades and three razor blades. As far as typological classification is concerned, blades are represented by two finishing blades, three preparation blades, of which two have original surface, and one comb blade. The little blades are documented by one specimen and two preparation blades of which one betrays remnants of the original surface of the mother rock.

Of the total number of twelve blades and razor blades, the base was preserved in the case of eleven specimens. The base indicates a way of the preparation of cores of which semi-products come, and informs about the technology of their brushing-off as well. The assemblage is dominated by blades with a primarily faceted base (three pieces), but in four cases there also occurred blades with a pointed base. On three specimens an untreated base was detected, and in one case a broken base with small dorsal reduction occurred. As far as the level of preservation of the original surface in the case of blade and razor blade artefacts is concerned, out of the twelve artefacts it was preserved only on three pieces. Since the artefacts are acquired from the mining pit - pingo for rádiolarite, it is assumed that in the place itself there was no mining of finishing blades and small blades. Or at least not to the extent as regards the typological composition of find assemblages of chipped industry from settlements. One may perhaps rightfully assume that in the area of the pingo there was only preparation of proto-cores and cores, which were dispatched to the close surrounding and from there probably distributed further. Such assumption is indicated by the preparation blades and razor blades with preserved original surface.

The category of flakes and waste includes 92 artefacts. Of the total number as many as 82 specimens were not retouched, for the utility retouching was detected only with eight flakes. Two artefacts were assessed as waste, of which one was a scale.

As for the typological representation, preparation flakes (51 pieces) include the following: 11 specimens belong to massive preparation flakes; 37 specimens belong to preparation flakes; in three cases it is a preparation flake with lateral scale. The assemblage contained also two reparation flakes and a reparation flake from the silicite of glacigene sediments (Fig. 2, n°12), as well as trimming flakes which originated during the treatment of core in the phase when a technical error during the acquisition of semi-products occurred. In four cases the base was not preserved or it could not be determined. Preparation flakes were dominated by the untreated base which could be observed in as many as thirty-one cases. On the contrary, the pointed base was determined in eleven cases. Moreover, on five artefacts the base was primarily faceted, on two it was adjusted by several hits, and on one specimen the base was broken and one flat. On reparation flakes the base was untreated or pointed. Into the group of flakes bearing no traces of the original scale were included thirty-two artefacts. Twenty-one of them had an untreated base, while in five cases the base was pointed. Only in one case the primarily faceted base occurred, as well as the base treated by several hits. In addition to them, the assemblage contained three flakes with scale of which two artefacts had untreated base and one base was pointed.

A level of preservation of the original surface was observed in the group of flakes and waste as well. A relatively numerous group (up to 26 artefacts) showed a differently preserved original surface, and a group of flakes with partially preserved surface was confirmed in fifteen cases. A relatively numerous group (47 pieces) was made up also by artefacts with negative surface.

The category of instruments is minimally represented. Based on the evaluation, a total of five artefacts was distinguished. Four flake semi-products were used for their making and in one case a remnant of the raw material was used to make the instrument. The group of scrapers is represented by two pieces, including an indistinctive flake scraper as well as a simple flake scraper. On the contrary, to the group of gravers it was possible to include just one specimen, which was made on the preparation flake. Of other instruments, one massive and one indistinctive side scarper appeared.

As it was mentioned in the introduction, during the last years mining fields for silicite raw material were identified in western Slovakia. The discovery of a surface pit in Bolešov, location Pri troch kopcoch, and subsequently also the area with six pits in Krivoklát, location Bukovina, indicate the strategic territory within the klippen belt of the White Carpathians. Both sites are situated in the 1900 m air line distance. It may be rightfully expected that the area played a significant role in supplying the Neolithic and Eneolithic population with the raw material for the production of chipped industry. Further informations about this region were decribed in the article in periodical Slovenská archeológia (Cheben - Cheben, 2010). Results of the exploration of the pingo in Vršatské Podhradie permits one to clearly state that it is a prehistoric mining work originated by the acquisition of radiolarite in a mining way. The dating may be made more precise by analysis of wooden charcoals from a layer in pingo I. The analysis of a non-numerous assemblage of chipped industry does not so far permit a more exact chronological or cultural classification. The technique of the production of artefacts indicates that radiolarite was most probably mined in the course of the Neolithic. However, it cannot be ruled out that further detailed exploration of the observed territory will increase the number of mining fields registered so far. The field research in the locations identified so far continues to have a significant role, to clearly determine that they are prehistoric mining works, and based on the acquired inventory, to chronologically classify them.

## References


Bárta, J. 1961. K problematike paleolitu Bielych Karpát, Slovenská archeológia 9, 9–28.

Bárta, J. 1979. K problematike proveniencie surovín na výrobu štiepanej industrie v paleolite Slovenska, Slovenská archeológia 27, 5–15.

Dzieduszycka-Machnikowa, A., Lech, J. 1976. Neolityczne zespoły pracowniane z kopalni krzemienia w Sąspowie. Wrocław – Warszawa – Kraków – Gdańsk.

Cheben, I., Illášová, Ľ., Hromada, J., Ožvoldová, L., Pavelčík, J. 1995. Eine Oberflächengrube zur Förderung von Radiolarit in Bolešov, Slovenská archeológia 43, 185–204.

Cheben, I., Cheben, M. 2010. Research on radiolarites of the White Carpathian Klippen belt, Slovenská archeológia 58, 13-52.

Mišík, M. 1999. Príspevok k litológii a paleografii rádiolaritov zo Západných Karpát, Mineralia Slovaca 31, 491–506.

Oliva, M. 1998. Gravettien východní Moravy, Časopis Moravského muzea – Vědy společenské 83, 3–65.

Vencl, S. 1967. K otázce datování tzv. vlárskeho paleolitu, Musaica (Sborník filosofické fakulty univerzity Komenského) 7, 3–13.

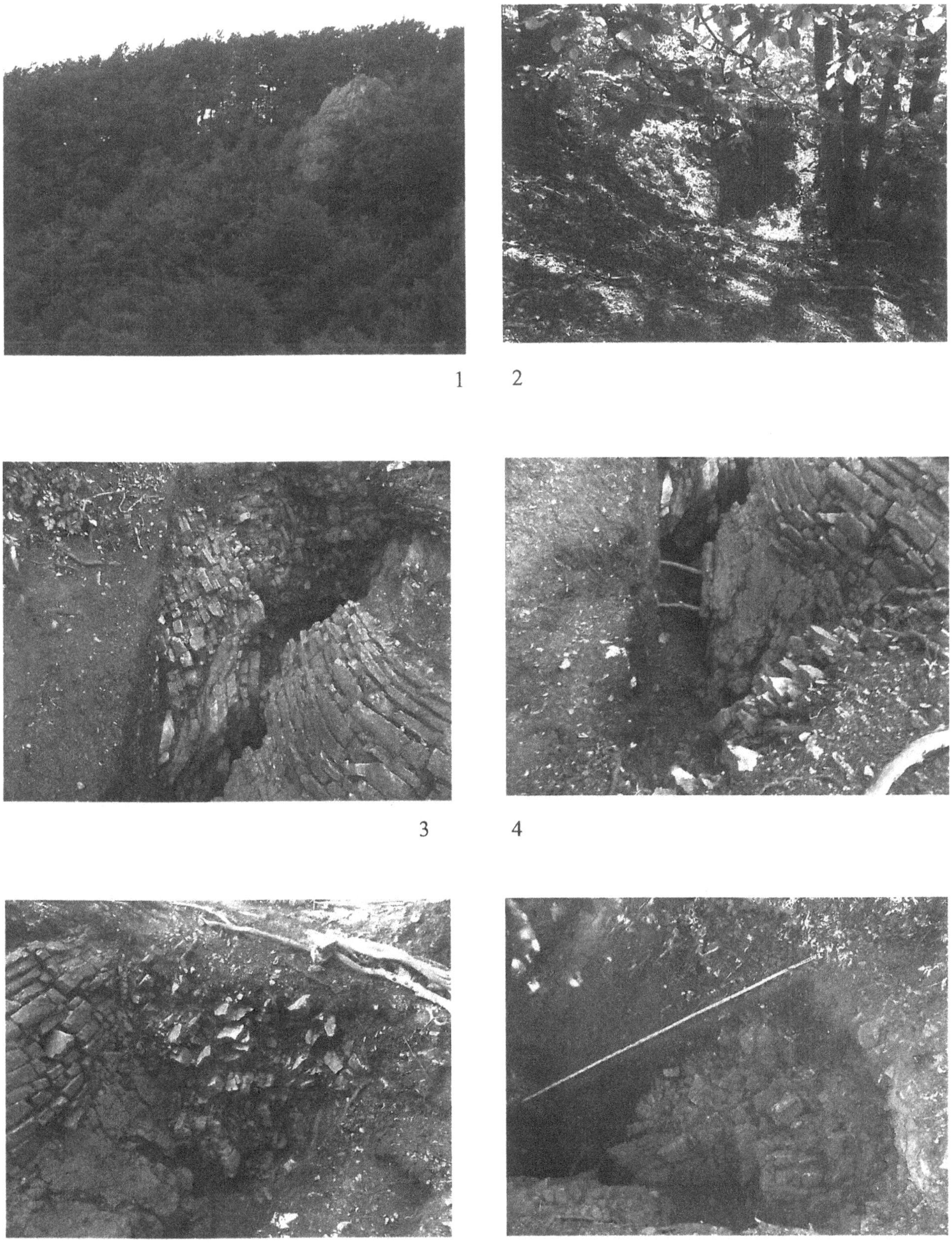


Fig. 1 : 1. hill area above Horná lysá, 2. cauldron-shaped sunk pit, 3. eastern wall with the layer of radiolarite, 4. lower part of the pingo, 5. southern part of pingo II, backfill made up of fragments of radiolarite and broken limestone, 6. southern part of pingo II, backfill made up of fragments of radiolarite and broken limestone.

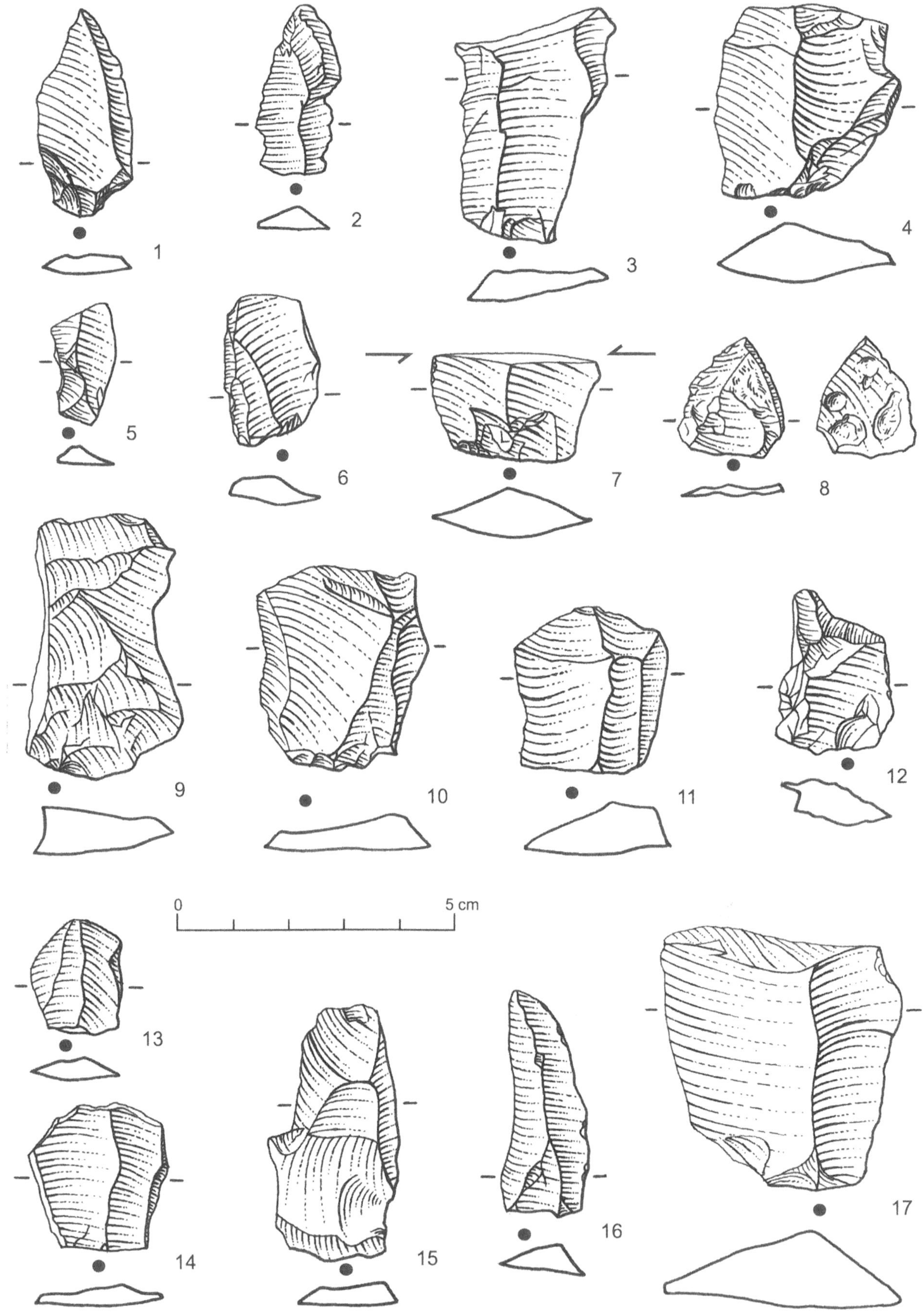


Fig. 2 : Chipped stone industry from the pinga.

# Features of flint mining and processing during the Chacolithic period in the Southeast of Europe (based on the materials of the Tripolian Bodaki settlement, Ukraine)

Natalia SKAKUN, Anaik SAMZUN, Boryana MATEVA & Vera TEREKHINA

## Abstract

After a short presentation of the Tripolie Culture (Chalcolithic period), the authors focus this article on the specificities of the lithic production on the site of Bodaki (Western Ukraine, 4th Mil. B.C.). The lithic production is linked to the outcrops of flint situated very closed to the site. The Volhynian flint is a of high quality and allows the knapping of large blades probably made with the help of crutches whose several exemplaries were found in situ. Moreover, a few large hips of lithic production have been found close to the flint settlement as well as on the site itself closed to the dwelling structures. They correspond to open air lithic workshops and contained large quantities of cores, flakes, stone hammers, retouchers, blades as well as various tools. Use-wear analysis indicate that a large part of them were never used. Thus, the site of Bodaki can be considered as a center of lithic production where a community of craftmen lived here, exploited the raw material and produced flint blades and tolls which circulated sometimes on long distances since a few artefacts made of Volhynian flint were found in other parts of Ukraine but also in Poland and Hungary.

**Keywords:** flint workshop; Tripolje; Ukraine; Chalcolithic

The flint mining methods, the processing techniques of it in the workshops and the characteristics of the flint processing organization into the system of ancient economy — all these subjects are the most discussed in the Stone Age archaeology. In the archaeological literature, including Russian literature, quite a few diverse works are devoted to this problematic. They include the analysis of the archaeological context, the technical-morphological, experimental and use-wear studies of lithic materials and also results of the other natural sciences (geology, petrography, etc.) (Gurina 1976; Krizhevskaja 1960; Mirsaatov 1973; Kasymov 1972; Comşa 1976; Kunchev & Nachev 1984; Matjuhin 1994, 1995; Semenov 1957; Bibikov 1965; 1966; Chernysh 1967; Popova 1980; Skakun 2006; Petrun' 1967; 1971; 2012; Kovnurko 1962, 1975; Kunchev *et al.* 1981).

This problematic is urgent for the Palaeometal epoch cultures too, since in many European territories, in spite of discovery of the copper existence, flint continued to remain the most used raw material for tools (Skakun 1981; 1984; 1996a, b; 2006; Sirakov & Tsonev 1995, 2001). From the point of view of socio-economic development, even in such progress regions of the Chalcolithic world in the Ancient East and the South-eastern Europe, the flint artefacts were still used everywhere and successfully competed with a metal in many economic sectors. Unfortunately, the due attention isn't always paid to the analysis of this type of archaeological data, and often flint implements of this time are considered as archaic, connected with former Neolithic period that doesn't correspond to a type of used raw material, especially techniques of flint knapping.

In the Chalcolithic period in South-eastern Europe characterized by the flourishing of agricultural cultures (V-IV$^{e}$ Mil. B.C.) it was begun largely exploitation of big nodules Cretaceous flint deposits, which were rarely used during former periods. In North-Eastern Bulgaria, during the Chalcolithic, the richest Dobrudjian flint deposits are developed and there arise new processing techniques of this raw material (Skakun 1981; 1984; 1986; 2008; Nachev, 2009; Gurova & Nachev 2008; Gurova & Bonsall 2010; Mateva 2008; Mateva 2009; 2011; 2012).

The technology of flint knapping of this period made possible to make high quality preforms in its technical parameters — long regular blades. Their fragments due to very sharp edges often used as a variety of tools and inserts without additional processing (Skakun 2008). Experimental analysis corroborate that the most qualified product are characterised by a right profile, sharp symmetrical edges, a small butt and were probably made by pressure with a lever (Pelegrin 2006; 2012). The mass of flint tools was probably produced by specialists-i.e. professionals knappers in workshops and settlements, situated near the sources of raw material which circulated on the whole territory of Bulgaria and even out of its territory for instance to the Lower Danube river in Ukraine. Noteworthy is the fact that at the same time with the settlements with workshops there were settlements where main part of tools was made from Dobrujian flint large blades. There are discovered treasures of finished blades, and any object connected with flint processing: cores, waste products, knapping tools, etc. are absent. These settlements, such as sites of Varna archaeological culture in Bulgaria's Black Sea coast were consumers of finished products (Skakun 2008).

During the Chalcolithic period, a bit later than in Bulgaria, in the area of Kukuteni-Tripolje tribal community (Rumania, Moldavia, Ukraine), began the expansion to North-Western Ukraine (Popova & Chernysh 1967). Penetration of the Tripolian groups on this territory was directly linked to exploration and exploitation of big nodules Cretaceous flint deposits at the Volhynian Elevation. Close to the sources, workshops (Konoplja 1982) and workshop sites, i.e. settlements which were specialized on the making of tools, during this period appeared (Skakun & Samzun 2004; Skakun 2004; 2005).

Now one of the most studied Tripolian site-workshop is the Bodaki settlement (Cynkałowsky 1969; Skakun, 1996b; Skakun 2004; Skakun et al., 2005). It is situated on the left bank, on a terrace of the Horyn river which flows into the Pripiat' river — a tributary of the Dnieper river (Fig. 1). The environs of the site, gullies and dry streams bed are totally scattered with flint nodules of various forms and their length can reach 50 cm (Fig. 2). The numerous flint deposits are laid closed to the surface and it was possible at this time to get them in open air or in intentionally digging mines. Two deposits are located at a closed proximity of the limit of the Bodaki site. One is located at about 1 km from its Eastern limit of the site and the other one is located at about 100 m from its Western limit. The flint of these sources shows a fine aspect, homogeneous in fracture, without cracks, seldom, with extraneous inclusions that complicate knapping. It is generally black colour, opaque or translucent, with sometimes-veined grey concretion of various forms and dimensions (Fig. 3). The cortex is dense, fine (thickness less than 0,1 cm). The Volhynian flint, as well as Donetsk's one from the right bank of the Dnieper river is considered one of the best quality in Ukraine.

The Bodaki site is a complex of dwelling structures and excavated domestic pits surrounded on three sides by a ditch in a horse shoe form (Fig. 4). As result of many years of investigation on the settlement and around it were found different in purpose sites related to flint processing.

Places of the preliminary stage of flint processing have been found outside the settlement, at its western limit, near the ravine, at its slopes and bottom there are many nodules various shapes and sizes. Some of them are raw material, other are tested, there are also precores and a large quantity of shaping out flakes and cortical flakes.

In the central part of the site, where the remains of built structures are not numerous, in a large 8 shaped pit-dwelling, an area of about 20 m$^2$ was found a flint workshop. Contiguous to one of the walls of the pit-dwelling, there was a carefully built subrectangular fireplace. This is single structure such this type on the site.

In the pit-dwelling consisting of two levels, there was more than 1500 flint artefacts. Among them, 52 cores, 11 spheric percuteurs, 2 punch made of deer antler, 3 retouchers on fractured blades, long blades and their fragments and also tools (end scrapers, burins, borers, drills, knives on blades with retouch on distal part) and a large quantity of flakes of various lengths, sometimes cortical (Fig. 5). Use-wear analysis showed that except the instruments used for knapping, a part of the tools unearthed in this workshop, had not any trace of utilisation.

Very closely to the workshop, 2 refuse pits contained mass of wastes of production: shaping out flakes, flakes from preparation of striking platform, chips been used since is very big, but some of them were well retouched by pressure. It is possible that they correspond to testing or apprentice works.

The majority of the cores are oblong shaped and their dimensions are rather large (20 cm — 12×15 cm — 10×8 cm — 6 cm). Observing their negative, the preparation of their knapping is oriented to the making of the most long blades as possible (length: 15 cm; largeness: 1,8-3 cm), though only a few blades of such length were found on the site (Fig. 5).

Besides the specialised flint workshop, between the pit-dwellings area with large concentrations of waste productions was fixed. The excavation showed that in ancient period the slope of the terrace on the river has been levelled with the help of filling which was added, packed and intentional platform of about 40 m$^2$ has thus been set up (Fig. 6).

Large blocks were found on it, which have may be maintained a building with posts with a light roof. In this area very large quantities of flakes and debris and more than 2000 artefacts, including 36 cores and their waste of knapping, 250 blades, including 44 complete ones and 206 fragments, 1576 flakes of various sizes (691 cortical), 137 tools including 6 hammerstones on spherical flint nodules and a large number of debris and wastes.

Specific tools and archeological context of this place clearly indicate that it was a flint workshop in open air.

In the dwelling part of the occupation where are set pise buildings and pit-dwellings, the artefacts related with flint processing are not so numerous. For instance, study of materials of one of the dwellings showed that 211 flint artefacts including 63 tools with traces of use and also a compact cluster of 12 unbroken blades probably kept in a small bag as blanks. The tools for knapping are here absent, cores are found two, and other materials are represented by flakes (Skakun & Mateva 2011).

The results attest that at Bodaki, the first stage of the flint processing takes place outside of the site itself whereas the knapping of the cores and preparation of the tools are mostly done in large workshop and in open air area. To a lesser extent these activities can be done in the proximity of the dwelling structures and inside them. The finding of a large number of cores and waste of production, large quantities of tools including many unused, allow to say that the Bodaki people were specialized on flint processing and in particular in the production of large blades and their tools which spread inside Tripolje Culture areas and also into the other territories. The quality and the standardisation of their production of large blades confirm that it was the work specialized craftsmen worked on them. The analysis of flint materials shows the different stages of the chain operations: the cores (the implements are flint hammerstones, horn punches, flint retouchers), wastes of production and the products of knapping themselves (large blades and their tools) allow to confirm that at Bodaki were adopted methods of work including knapping with the help of a deer antler punch and may be lever device in order to get more efficient and qualified products. The more realistic reconstruction of it was proposed by J. Pelegrin (2012).

In the economy of this site, looking at the archaeological context and the rich assortment of the used tools found the dwelling structures and domestic pits, with the evidence of the specialisation of the flint processing, all these aspects reflects specifically aspects of the economy of the Chalcolithic period, i.e., the appearance of sites which play the role of production centres and yield what it is necessary and which is distributed.

The main way for food acquisition were cultivation of plants and animal breeding and work of skin and leather, wood, antler and horn, bone, mineral colours were also produced. Moreover, on the site there was pottery workshop. Its quality of ceramic such as in central Tripolian contemporaneous sites (Fig. 7).

Excavations at Bodaki also provide with interesting aspects concerning sacral aspects during the Chalcolithic period. On the floor of the workshop a spot of ochre surrounded of small wallet made in clay was has been exposed in one of its corners, were found cores.

On the working place small burned pise structure was excavated and in the proximity, a few flint blanks covered with ochre. On the floor of pit-dwelling located near the working place was found ochre and closely a bull's skull was excavated (Fig. 8).

In the other pit-dwelling, a large stone like the bull's head was found and completed by 2 tusks of wild boar may be imitating the horns of the bull. Between them were found 2 large flint end-scrapers. The «model» was covered with ochre. These observations allow to consider the existence of ritual which could be related with the flint processing, indicating the expansion of a mythology among Tripolian tribes.

In the actual archaeological literature, the discoveries of flint artefacts from Volhynian flint are often mentioned. They were used not only on the Tripolian sites in Ukraine and Moldavia but were also spread on the contemporaneous sites of Poland and Hungry (Fig. 9).

It is worthnoting that large blades from local flint are attested in the burials of the ancient nomad tribes of the steps from the left bank of the Dnieper river who had contacts with the contemporaneous farmer culture (Skakun 2006).

Thus, the study of the material of this South-Eastern part of Europe shows that the appearance of flint production centres during the Palaeometal period was not an isolated phenomenon (Skakun 1984, 2002, 2005, Tsvek 2005).

The high level of the technology of the flint production are characterized its general aspects including development of the large nodules of Cretaceous flint, the appearance of flint workshop centres with specialized and professional knappers having a high level of treatment technology of raw material and devoted to the production of large blades which have circulated on rather long distances.

These data allow to consider that the flint processing of the Chalcolithic period occurred and replaced the domestic production of the earlier phases of the Stone Age and became one of the most evident example of the appearance of early common craftsmanship.

## References

Bibikov, S.N., 1965. Drevnie kremnevye vyrabotki na gore Beloj v rajone Kamenets-Podol'ska (Ancient flint productions in mount Bely in the Kamenets-Podolsk district). In: *Materialy sessii, posvjaschennoj itogam arkheologicheskih i jetnograficheskikh issledovanij v 1964 g. v SSSR. Abstracts*. Baku, 56-58.

Bibikov, S.N., 1966. Drevnie kremnevye vyrabotki v Srednem Podnestrov'e (Ancient flint productions in the Middle Lower Dniester river). In: *Sborník Národního muzea v Praze. Acta Musei Nationalis Pragae*. Vol. XX. N°1/2, 3-7.

Chernysh, E.K., 1967. Tripol'skie masterskie po obrabotke kremnja (The Tripolian Culture flint workshops). In: *Kratkie soobschenija Instituta arkheologii*. Moscow. Vol. 111, 60-66.

Comşa, E., 1976. Les matières premières en usage chez les hommes Néolithiques de l'actuel territoire Roumain. In: *Acta Archaelogica Carpathica*, XVI. Kraków, 239-249.

Cynkaşłowsky, A., 1969. Osiedle kultury trypolskiej w Bodakach nad Horyniem (Bodaki, the site of Tripolie Culture on the Horyn river). In: *Wiedomości archeologiczne*. No. 34, 221-227.

Gurina, N.N., 1976. *Drevnie kremnedobyvajuschie shakhty (Ancient flint mines)*. Leningrad. 180 p.

Gurova, M. and Nachev, Ch., 2008. Formal Early Neolithic Flint Toolkits: Archaeological and Sedimentological Aspects. In: R.I., Kostov, B., Gaydarska, M., Gurova (eds.). *Geoarchaeology and Archaeomineralogy. Proceedings of the International Conference, 29-30 October 2008*. Sofia, 29-35.

Gurova, M. and Bonsall, C., 2010. 'Balkan flint' in south-eastern European prehistory: chronological, cultural and social perspectives. In: *The European Archaeologist*. No. 32, 30.

Kasimov, M.R., 1972. *Kamneobrabatyvajuschie masterskie i shakhty kamennogo veka Srednej Asii* (Lithic workshops of the Stone Age in Central Asia). Tashkent: 159 p.

Konoplja, V.M., 1982. Obrobka kremenju naselennjam Zakhidnoj Volini za doby midi-rannoj Bronzy (Flint processing of the Western Volhynian people of the Chalcolithic — Early-Middle Bronze Age). In: *Arkheologija*. No. 37. Kiev, 17-31.

Kovnurko, G.M., 1962. K izucheniju svoistv kremnja (Study of the flint property). In: *Kratkie soobschenija Instituta arkheologii*. Moscow. Vol. 92, 97-99.

Kovnurko, G.M., 1975. Petroarkheologicheskie issledovanija kremnja Russkoj platformy (Petroarchaeological flint studies of the Russian platform). In: *Folia*. Vol. 16. No. 10, 43-47.

Krizhevskaja, L.Ja., 1960. Kremneobrabatyvajushhaja masterskaja i poselenie na severo-vostoke Bashkirii. In: *Paleolit i neolit SSSR. T. 4. Materialy i issledovanija po arheologii SSSR*. No. 79, 239-281.

Kunchev, K., Nachev, Iv., Kovnurko, G., 1981. Krem'chnite skali v Bulgaria i tjakhnata eksploatatsija (Flint sources in Bulgaria and their exploitation). In: *Interdisciplinarni izsledvanija*. Vol. VII-VIII, 41-59.

Kunchev, K.S. and Nachev, I.K., 1984. Aptian and Cvaternery flint in East Bulgaria. In: *IIIrd Seminar on petroarchaeology, 27-30 August*. Plovdiv, 65-82.

Mateva, B.I., 2008. Eneolithic flint workshop-settlements in North-eastern. In: L., Longo and N., Skakun (eds.) *'Prehistoric Technology' 40 Years Later: Functional Studies and the Russian Legacy*. Oxford, BAR International Series 1783, 451-452.

Mateva B.I., 2009. K voprosu ob organizatsii pervichnoj obrabotki kremnja v jepokhu jeneolita (geologicheskie dannye i jetnograficheskie paralleli) (Problem of the organisation of preliminary flint processing during the Eneolithic period. Geological data and ethnographical parallels). In: *Bibikov S.N. i pervobytnaja arkheologija*. Sankt-Petersburg, 350-355.

Mateva, B.I., 2011. Exploiting of flint deposits in northeastern Bulgaria in Chalkolith. In: *The Lower*

*Danube in prehistory: landscape changes and human-environment interactions. Proceedings of the International Conference Alexandria, 3-5 November 2010*. Bucureşti, 173-179.

Mateva, B.I., 2012. L'extraction de silex en Bulgarie du nord-est à l'époque Chalcolithique. In: *Des grandes lames en silex dans toute l'Europe, Actes de la table-ronde international L'Europe, déjà, à la fin des temps préhistoriques, Tours (Indre-et-Loire, France), 7 Septembre 2007. 38 supplément à la Revue Archéologique du Centre de la France*, 85-90.

Matjukhin, A.E., 1994. Paleoliticheskie masterskie v bassejne nizhnego Dona. (Palaeolithic workshops in the basin of the Lower Don river). In: *Arkheologicheskie vesti*. No. 3. Saint-Petersburg, 25-37.

Matjukhin, A.E., 1995. K voprosu o proiskhozhdenii i prirode paleoliticheskikh masterskikh (Problem of the origin of the Palaeolithic workshops). In: *Arkheologicheskij al'manakh*. Donetsk. No. 4, 31-40.

Mirsaatov, T., 1973. *Drevnie shakhty Uchtuta (Ancient mines of Uchtut)*. Tashkent: 260 p.

Nachev, Sh.I., 2009. Osnovnite tipove flint v Bulgarija, kato surovini za naprava na artefakti. In: *Interdisciplinarni izsledvanija*, 7-22.

Pelegrin, J., 2006. Long blades technology in the Old World: an experimental approach and some archaeological results. In: J., Apel and K., Knutsson (eds.) *Skilled Production and Social Reproduction. SAU Stone Studies*. Upsalla. No. 2, 37-68.

Pelegrin, J., 2012. Grandes lames de l'Europe Néolithique et alentour (Large blades from Neolithic Europe and abroad. In: *«L'Europe, déjà, à la fin des temps préhistoriques Des grandes lames en silex dans toute l'Europe», Actes de la table-ronde international. Tours (Indre-et-Loire, France), 7 Septembre 2007*, 38e supplément à la Revue Archéologique du Centre de la France, 15-43.

Petruń, V.F., 1967. K petrograficheskomu opredeleniju sostava i rajonov dobychi mineral'nogo syr'ja rannezemledel'cheskimi plemenami Juga-Zapada SSSR (Petrographical identification of composition and mining sites of the mineral raw material by early farmer tribes of the South-West of Soviet Union). In: *Kratkie soobschenija Instituta arheologii. Moscow*. Vol. 111, 50-59.

Petruń, V.F., 1971. K petrograficheskoj kharakteristike materialov kamennykh orudij paleolita (Petrographical characteristics of the lithic tools of the Palaeolithic). In: *Paleolit i neolit. Materialy i issledovanija po arkheologii SSSR*, no. 173. Moscow, 282-297.

Petruń, V.F., 2012. Vstup do arkheologichnoi petrografii Ukraini (The introduction to archaeological petrography in Ukraine). In: *Chelovek v istorii i kul'ture*. Vol. 2. Odessa, 110-122.

Popova, T.A., 1980. Kremneobrabatyvajuschee proizvodstvo tripol'skikh plemen (po materialam Polivanova Jara) (Flint processing of Tripolian tribes, based on the material of Polivanov Jar). In: *Pervobytnaja arkheologija. poiski i nakhodki*. Kiev, 145-163.

Popova, T.A. and Chernysh, E.K., 1967. Tripol'skoe poselenie u s. Bodaki (The Tripolian site near Bodaki village). In: *Zapiski Odesskogo arkheologicheskogo obschestva*. T. II (35). Odessa, 173-179.

Sirakov, N. and Tsonev, T., 1995. Chipped-stone Assemblage of Hotnitsa-Vodopada (Eneolithic / Early Bronze Age Transition in Northern Bulgaria) and the Problem of the Earliest 'Steppe Invasion' in Balkans. In: *Préhistoire Européenne*, 7, 241-264.

Sirakov, N. and Tsonev, T., 2001. The Lithic Technology Tradition of the Late Eneolithic in Central South Bulgaria. In: *Studies in Honour of Prof. J.Kozlowski. Kraków*. Vol. VIII: pp. 62-71.

Skakun, N.N., 1981. Eksperimenty v ekspeditsii 'Dobrudzha-79' (Experimentation in the expedition 'Dobrudja-79'). In: *Interdisciplinarni izsledvanija. Sofija*. Vol. VII-VIII, 59-64.

Skakun, N.N., 1984. Kremneobrabatyvajuschee proizvodstvo v epokhu paleometalla Bolgarii (Flint processing during the Palaeometal period). In: *IIIrd Seminar on Petroarchaeology, 27-30 august*. Plovdiv, 83-92.

Skakun, N.N., 1996a. K vorposu o kremneobrabatyvajuschem proizvodsve epokhi eneolita v Jugo-Vostochnoj Evrope (po materialam Bolgarii) (Problem of the flint processing during the Eneolithic period in South-Eastern Europe, based on material of Bulgaria). In: *Arheologija*. No. 3. Kiev, 124-128.

Skakun, N.N., 1996b. Le rôle et l'importance du silex dans le Chalcolithique de sud-est de l'Europe (sur la base du matériel provenant des fouilles du campement de Bodaki). In: *La Préhistoire au Quotidien*. Grenoble, 223-235.

Skakun, N.N., 2002. La production des grandes lames et outils en Ukraine au cours de la période énéolithique (culture de Tripolje): L'exemple du site de Bodaki. In: *Cahier des thèmes transversaux Archéologies et sciences de l'Antiquité*. III. Paris, 69-74.

Skakun, N.N., 2004. Predvaritel'nye resultaty izuchenija materialov tripol'skogo poselenia Bodaki (kremneobrabotyvajuschie kompleksy) (Preliminary results of the study of the material from the Tripolian Bodaki site (flint processing complexes)). In: *Orudija truda i sistemy zhizneobespechenija naselenija Evrazii (po materialam epokh paleolita — bronzy)*. Saint-Petersburg, 57-79.

Skakun, N.N., 2005. Bodaki — odin iz tsentrov kremneobrabatyvajuschego proizvodstva na Volyni (Bodaki is one of the centres of flint processing in Volhynia). In: N.N., Skakun, E.V., Tsvek, V.A., Kruts et al., *Arheologicheskie issledovanija tripol'skogo poselenija Bodaki v 2005, g.* Kiev; Sankt-Petersburg,64-79.

Skakun, N.N., 2006. Orudia truda i khozjajstvo drevnezemledel'cheskikh plemen Jugo-vostochnoj Evropy v epokhu eneolita (po materialam kultury Varna) (Tools and economy of the Eneolithic farmers of South-eastern Europe (based on the materials of Varna Culture)). *Trudy IIMK RAN*, XXI. Saint-Petersburg: 224 p.

Skakun, N.N., 2008. Les grandes lames de silex du mobilier funéraire des proto-éleveurs du sud de l'Europe orientale. In: *Préhistoire Anthropologie Méditerranéennes*, 14, 103-118.

Skakun, N.N. and Samzun, A. 2004. Preliminary results of the investigation of the Eneolithic site of Bodaki (Western Ukraine) with lithic workshops. In: *Acts of the XIVth UISPP Congress, University of Liège, Belgium, 2-8 September 2001*. Oxford, BAR International Series 1303, 225-229.

Skakun, N.N. and Mateva, B., 2011. Tipologicheski i funkcionalen analiz na krem'chen ansamb'l ot zhilische n° 17 na tripolskoto selishhe Bodaki, Zapadna Ukrajna (Typological and functional analysis of the flint complex of the dwelling structure n° 17 on the Tripolian Bodaki site, Western Ukraine. In: *Zlatnoto peto hiljadoletie. Trakija i s'sednite rajoni prez kamenno-mednata epoha: Dokladi ot mezhdunarodnija simpozium v Pazardzhik, Jundola, 26-30.10.2009 g.* Sofia, 287-295.

Tsvek E.V., 2005. Tsentry kremnevoj industrii plemen tripol'skoj obschnosti v Pobuzh'e i Podneprov'e. In: N.N., Skakun, E.V., Tsvek, V.A., Kruts et al. *Archeologicheskie issledovanija tripol'skogo poselenija Bodaki v 2005 g.* Kiev; Sankt-Petersburg, 80-92.

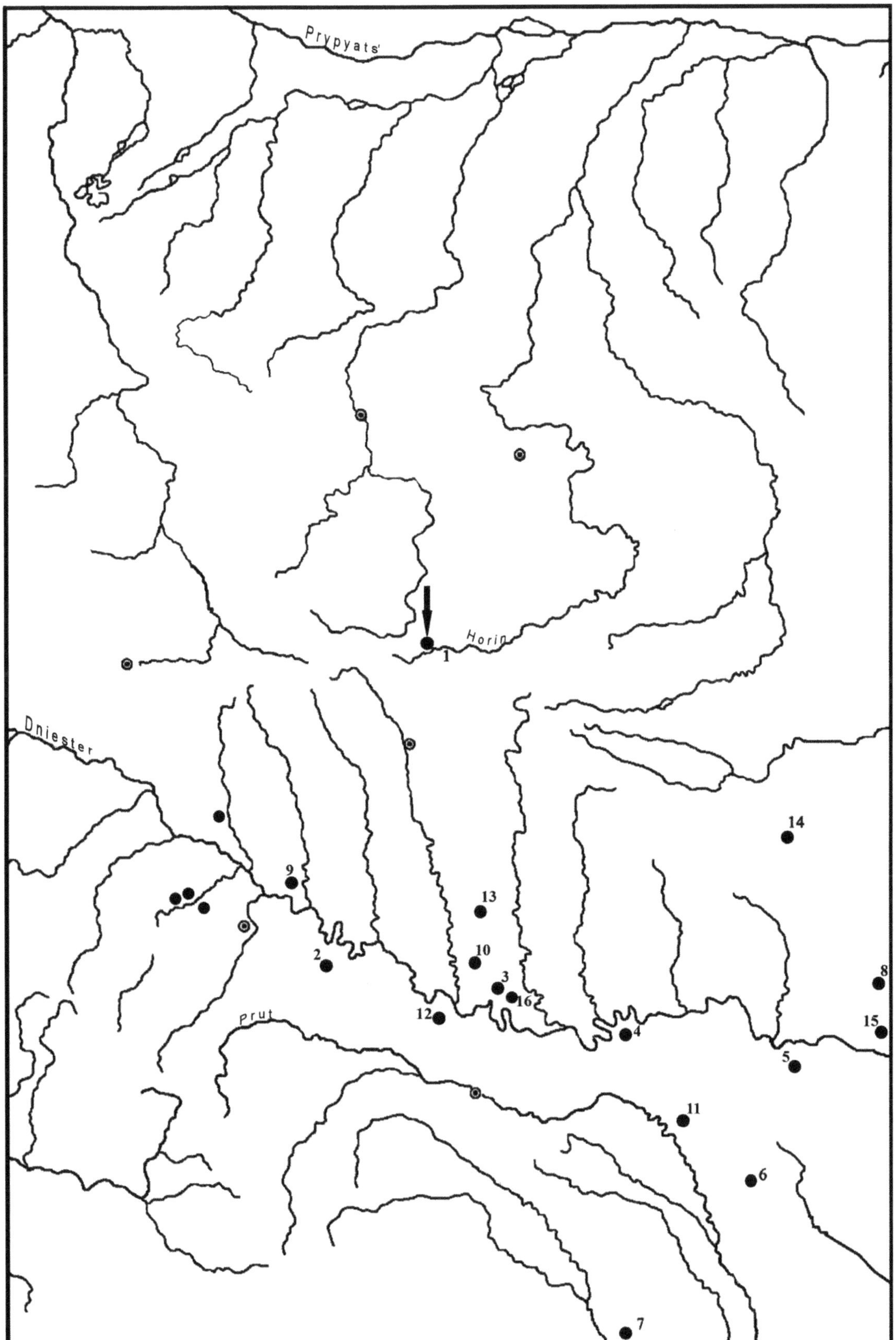


Fig. 1: Localisation of the Bodaki site and sites of the Tripolje BII: 1 – Bodaki; 2 – Neszvisko; 3 – Kudrintsa; 4 – Polivanov Yar II; 5 – Mereshovka-Chetetsuya III; 6 – Yablona; 7 – Kukuteni-Chetetsuya; 8 – Krinichki; 9 – Buchach; 10 – Gorodnitsa; 11 – Brynzeny VIII; 12 – Zaleschiki; 13 – Bil'che-Zolotoje; 14 – Nemirov; 15 – Rakovets; 16 – Zbruchanskoje.

Fig. 2 : Outcrops of the Volhynian flint on the Horyn river bank, to Western limit of the Bodaki site.

Fig. 3 : Volhynian flint nodules.

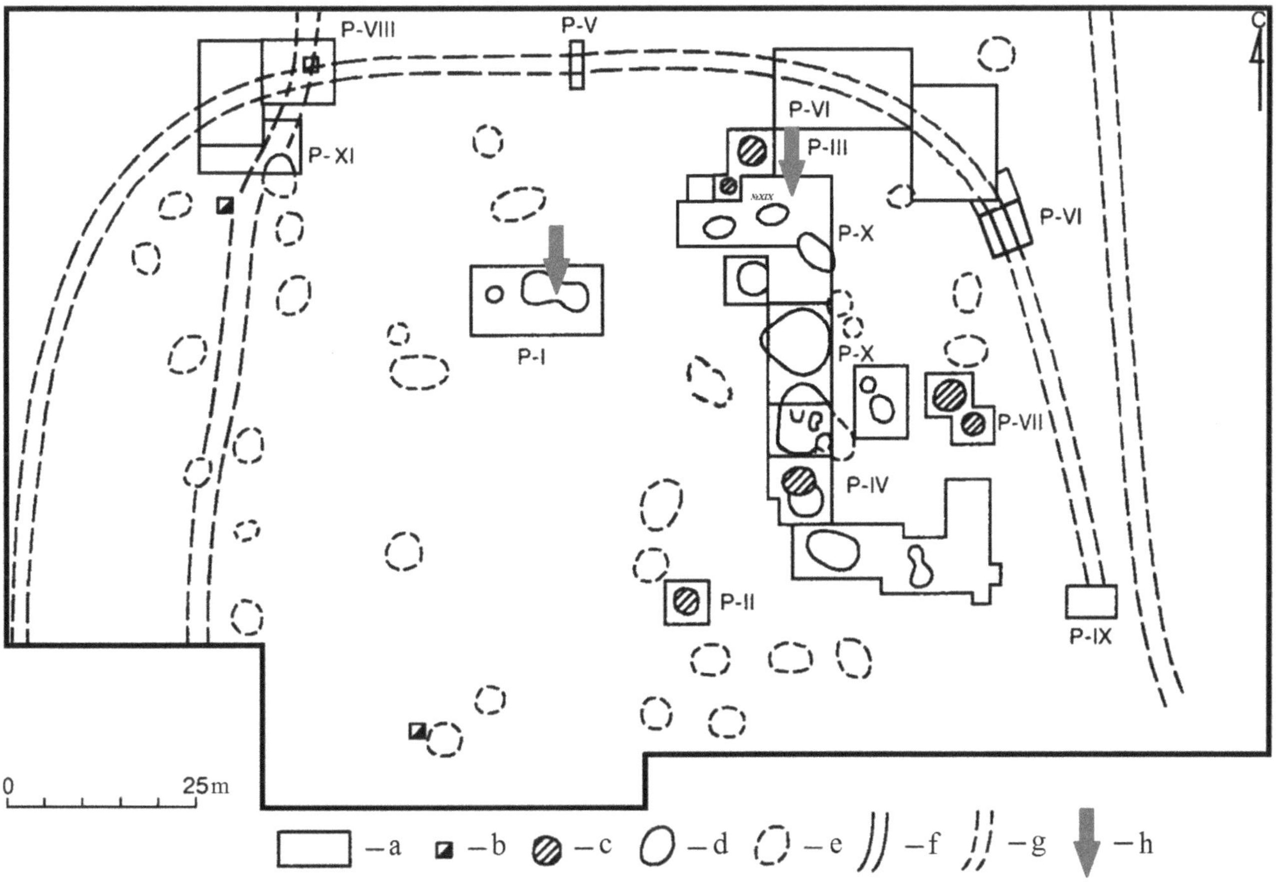


Fig. 4 : Scheme of the localisation of archaeological objects on the Bodaki site.
a – excavations; b – trenches; c – remains of pise buildings; d – excavated pit-dvellings and domestic pits; e – supposed pit-dwellings and domestic pits; f – excavated parts of the ditch; g – unexcavated parts of the ditch and objects of unknown function; h – flint workshop place on the site.

Fig. 6 : Bodaki: flint working place.

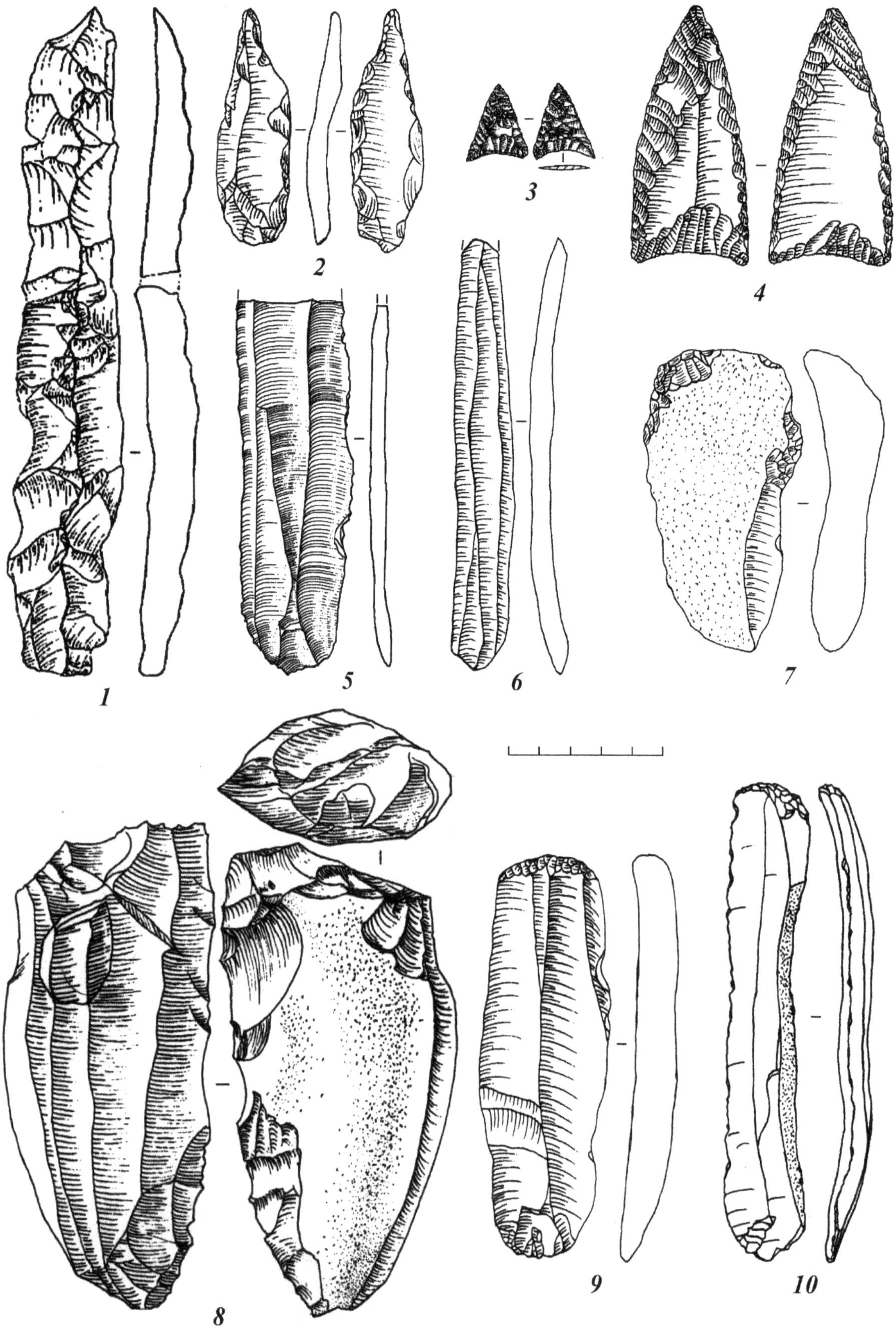


Fig. 5 : Flint artefacts of the Bodaki site.

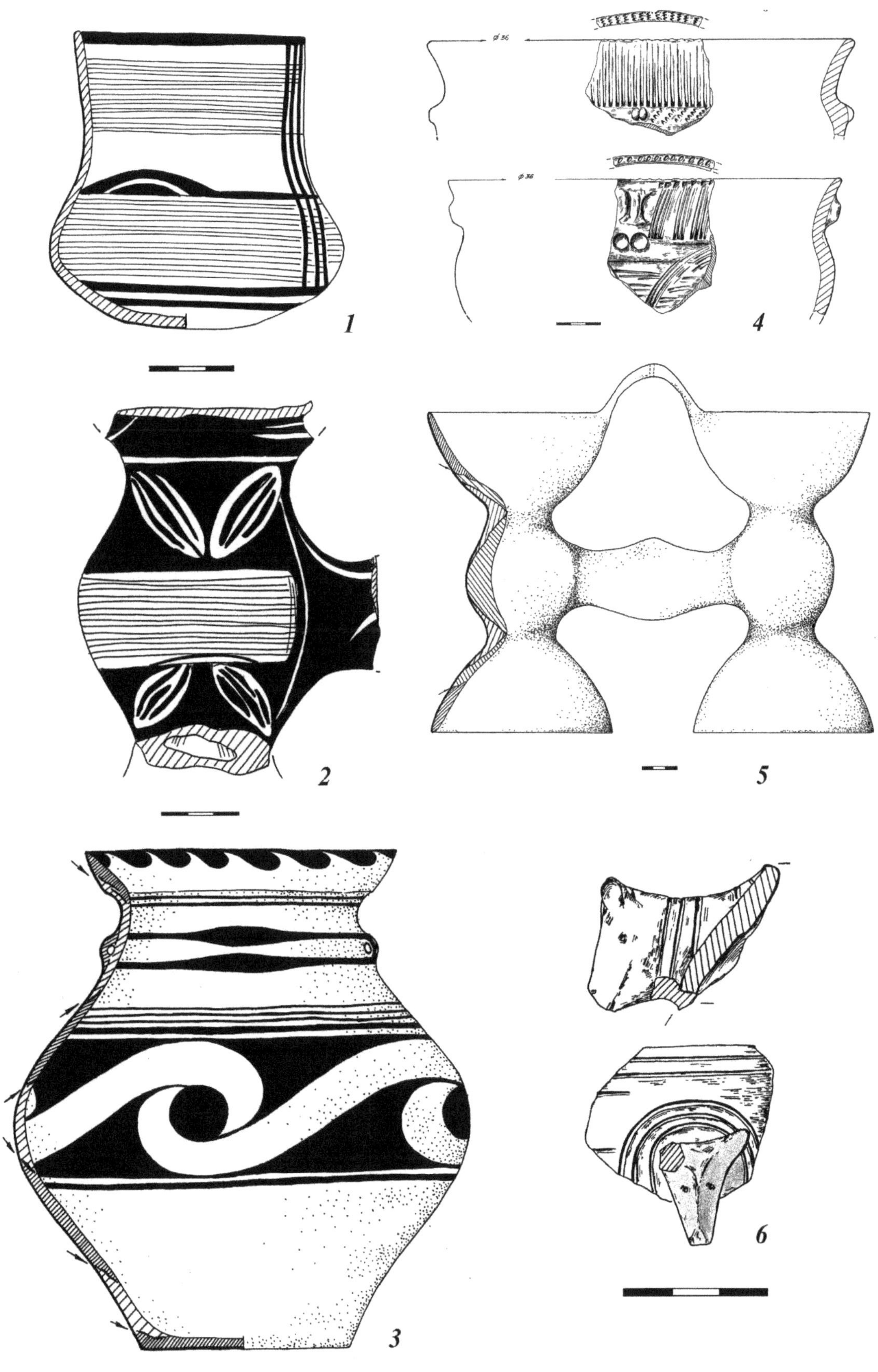


Fig. 7 : Bodaki: ceramics.

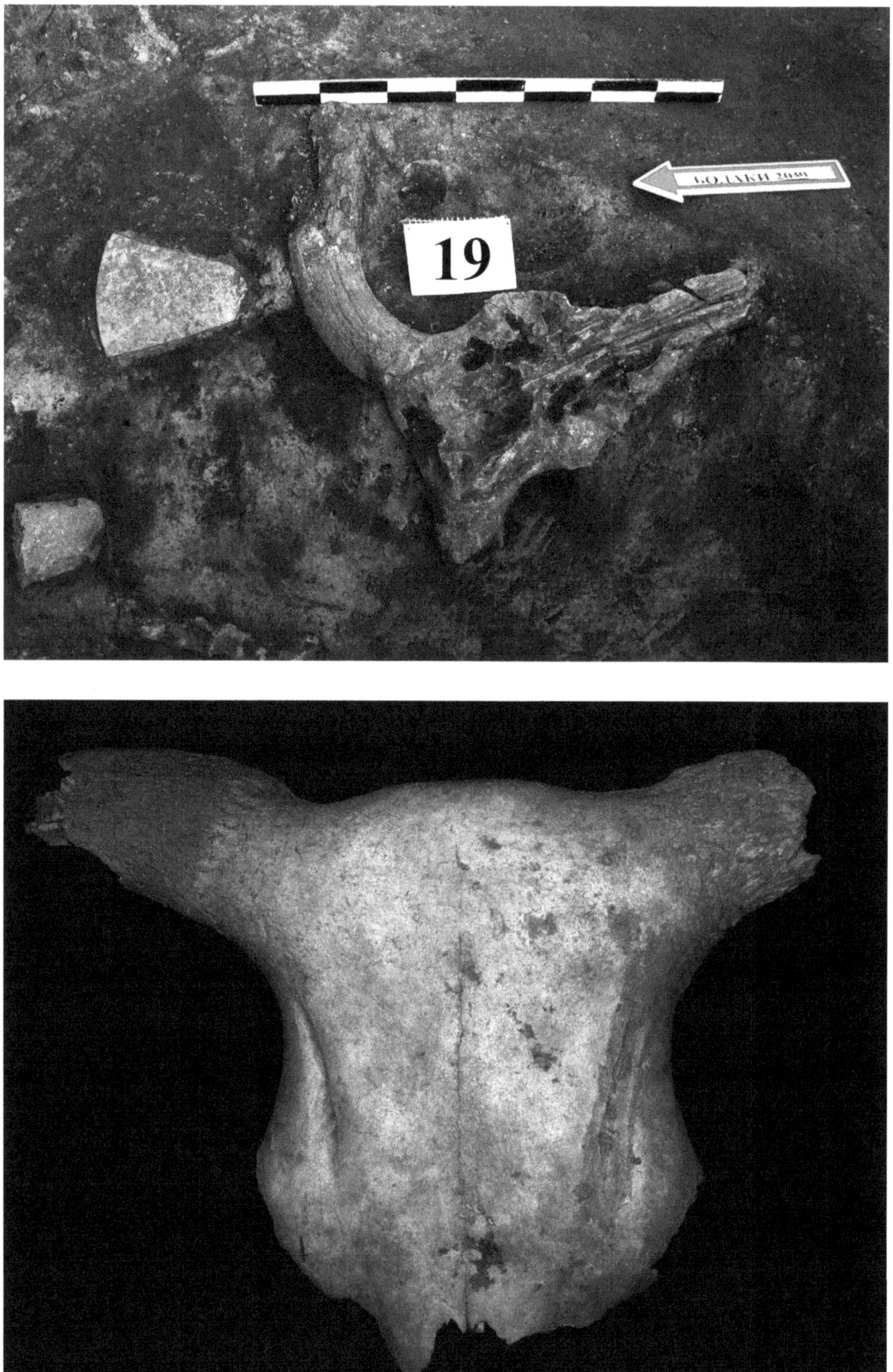


Fig. 8 : Skull of a bull from the pit-dwelling n° XIX.

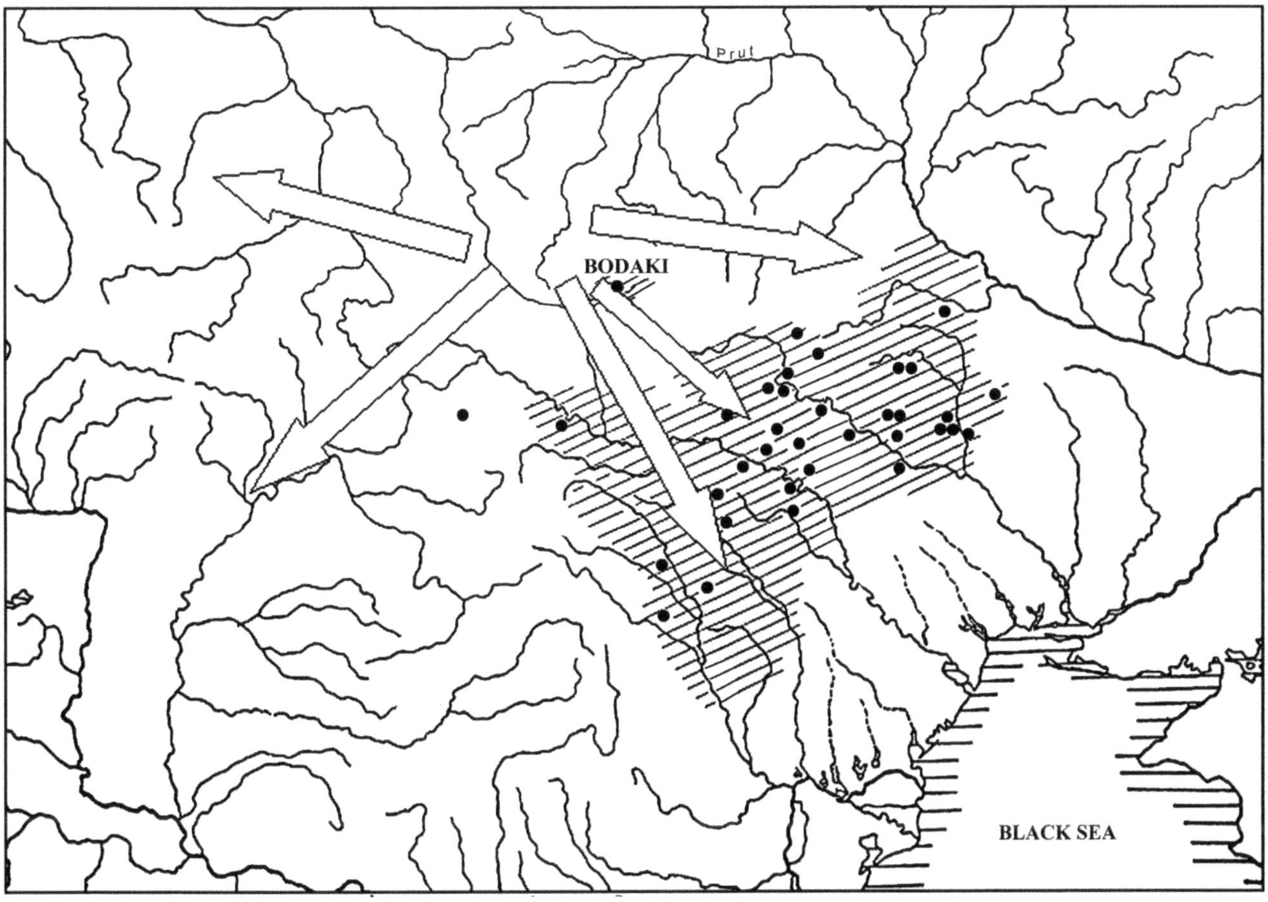


Fig. 9 : Map of spreading of the artefacts made from the Volhynian flint during the Chalcolithic period.

# Flint exchange during middle neolithic times – a case study from western Germany

Kathrin Nowak

## Abstract

In this study flint assemblages from twelve settlements located in the lower Rhineland and dating to the periods of Großgartach, Planig-Friedberg and Rössen are analysed. The main objective is to investigate if an exchange of flint artefacts occurred between those sites. While the flint of the type Rullen (Limburg, Belgian) represents the most important material during the Rössen period, the older flint assemblages display varying compositions of raw materials. In order to understand the mechanisms underlying the accommodation of flint, a factor analysis is applied to expose similarities and differences between the assemblages. Some sites are characterized by a high amount of waste products that occur during the manufacture of blades and flakes. Those settlements are very likely to provide flint products to other sites in the research area. An alternative thesis is suggested, according to which the sites with high production rates distributed flint into regions beyond the lower Rhineland. Both theses will be tested. In order to do this, the reduction sequences of the flint assemblages are analysed. Additionally the interregional distribution of flint materials is examined to uncover connections outside the study area.

**Keywords**: middle neolithic; flint exchange; factor analysis; distribution patterns.

## Introduction

The study area is situated in the lower Rhineland between Aachen and Cologne. Due to open cast lignite mining the region is well investigated, especially the eastern part of the Aldenhovener Platte. There and in neighbouring parts several excavated sites dating to the middle neolithic periods of Großgartacher/Planig-Friedberg (4900-4750 BC) and Rössen (4750-4600 BC) are located. According to the archaeological record, there existed either a very low population density or no occupation at all in the research area between the end of the early neolithic (Linearbandkeramik) and the middle neolithic. Thus, a new process of formation of settlements in the lower Rhineland at the beginning of the middle neolithic may be assumed (Zimmermann *et al.* 2006: 178-179). The settlements of the earlier periods under discussion, namely Großgartach and Planig-Friedberg, are indeed much less numerous than the ones dating to Rössen.

Twelve flint assemblages from excavated sites serve as the database in the following research. Descriptive analyses of this data have already been published providing the opportunity for further examination and interpretation (Gehlen and Schön 2007; 2009a; 2009b; Schwitalla 1997).

The main focus of this article lies on uncovering how the settlements were supplied with flint, in terms of whether in each settlement flint was procured or some sites were responsible for the procurement and provided other settlements with flint material. For this purpose a factor analysis is applied to detect differences between the assemblages in regard to the production of flint artefacts. Furthermore reduction sequences and the distribution of flint materials outside of the study area are under examination. In advance of those analyses an overview of the flint materials that occur in the middle neolithic assemblages is given.

## 1.Flint materials

Most of the flint materials found on neolithic sites located in the lower Rhineland can be identified as cretaceous flint belonging to the “Lanaye Member” and originating from the Belgian and Dutch Limburg (de Grooth 2011: 110-111). Traditionally, in neolithic research studies concerned with the study area a systematic distinction is made between two types of the Lanaye flint, the type Rullen and the type Rijckholt (Löhr et al. 1977: 162-163; Zimmermann 1988: 606-608). This approach provides advantages for the diachronic or synchronic comparison of flint assemblages. The type Rullen is characterized mainly by lighter colours including brown, yellow and light grey, whereas artefacts from the type Rijckholt display mostly a dark grey to light grey colour (Löhr et al. 1977: 156-157; Weiner 1997: 610). A detailed description of those types is newly given by M. Th. de Grooth (2011, 121-123), who also offers a further classification of the Lanaye flint. Both flint types are found in about 30 to 40 km distance to the research area. The outcrops of the type Rullen are located near Rullen and Aubel (Belgium), while the type Rijckholt originates a little farther to the north, in Dutch Limburg (Albers and Felder 1999; Löhr *et al.* 1977: 156-157).

In early neolithic times mainly the type Rijckholt was used (Zimmermann 1995). It is likely to have come from extraction sites at Banholt and not necessarily from the eponymous site at Ryckholt-St. Geertruid (de Grooth 2007: 167; Löhr *et al.* 1977: 156-157).

As shown in Fig. 1 both types are very common during the early middle neolithic. In this period only the site of Langweiler 10 stands out by a high amount of the type Rullen. (Gehlen and Schön 2009a: 242). However, considering the close proximity between Langweiler 10 and Langweiler 12 of 100 to 160 m, both sites might have belonged to a single settlement (Gehlen and Schön 2009a: 237). In that case, this settlement does not vary significantly from the others.

It can also be observed in Fig. 1 that the type Rullen represents the most important material in the Rössen period (Gehlen and Schön 2009b: 591).

Apart from the two types of Lanaye flint, other flint materials occur on the middle neolithic sites (Fig. 1). Among those are flint gravels, which are found locally and are of less quality compared to the other materials (Weiner 1997: 611-613). In addition, there are several flints from nearby or far away sources. Those have been arranged into two categories:

The category 1 is composed of flints that originate from outcrops more than 30 km away from the study area. The flints from this category are found to a considerable amount especially in the older assemblages. Among them is the silex d'Hesbaye and in high amounts the Obourg flint (Hainaut, Belgium). It has been suggested that they were obtained through contacts to the Villneuve-Saint-Germain/Bliquy culture (VSG/BQY) (Gehlen and Schön 2009b: 606). However, no Obourg flint is recorded from sites of this culture, although some of them are located in the Hainaut (Allard 2005: 172). According to M. Th. de Grooth (2011: 125) the Obourg flint is hard to distinguish from the Zeven Wegen type. She proposes to consider the Zeven Wegen flint as an alternative classification in cases like the one on hand. Small amounts of the Zeven Wegen can be found at the outcrops of Rullen (Albers and Felder 1999: 76), in much closer proximity to the Rhineland than the Obourg flint. Thus, at this point it must remain unclear, if relations to the VSG/BQY culture can be easily proposed in regard to the composition of the flint assemblages. Nevertheless contacts between this culture and the research area in the early middle neolithic are attested through pottery finds at the sites of Langweiler 10 and 12 (Spatz 1991).

The category 2 consists of flint types that have been obtained from nearby sources. Those are of minor importance in the flint assemblages under observation and shall not be discussed in more detail.

In summary, the early middle neolithic sites on one hand and the Rössen sites on the other hand display different compositions of raw materials. While several flint categories occur to a considerable amount in the older assemblages, the Rössen sites relied mostly on one flint source. This suggests the existence of different procurement strategies. It is probable that these are linked to the change in population density. During the older time period under discussion connections outside the research area have probably played a more important role than in Rössen. In times of low population density people had to search farther away from their local villages in order to find marriage partners or assistance in times of need (Peters 2011: 19; 26). In contrast to this, the main use of the yellow and brown flint of the type Rullen in Rössen could have served as means to emphasize a common identity within the research area and a distinction from other regions.

## 2. Factor analysis

As a multivariate, statistical technique the factor analysis is applied to detect interrelationships among variables (Hair *et al.* 2010: 104-105). In the here presented study, it is furthermore used to highlight similarities and differences between flint assemblages. Initially, this requires the selection of variables in relation to the research topic. Those are for example the amount of tools or waste products that offer indications on the production of flint artefacts. All variables are expressed as percentages. In advance of the analysis the variables were tested to assess their statistical adequacy. The ones with the best adequacy were selected for the analysis. They are shown in Table 1. A detailed description of this approach is given elsewhere (Nowak 2013).

Subsequently two factors were extracted from the five variables. A factor is defined by a set of variables that have a high loading on this factor (Hair *et al.* 2010: 129-130). Essentially the loadings represent correlations between the factors and the variables. The factor loadings of each factor are shown in the adjoined table (Table 1). They are additionally displayed in a diagram, in which the factors represent the axes (Fig. 2).

The first factor has a high negative correlation with a high amount of flakes without modifications. Those are basically the main waste products of flint manufacture. There is also a strong positive correlation between the first factor and a high amount of tools made of blades as well as tools made of flakes. The first factor can therefore be defined as a degree for the intensity of flint production. On the negative pole of the factor a high rate of production is displayed and in contrast the positive pole indicates a low production rate.

The interpretation of the second factor is less evident. There is also a considerable positive loading of a high amount of tools made of flakes. However the second factor is most strongly correlated to cores without modifications, meaning that they were only used for reduction and not additionally for example as hammer stones. Since those cores are generally rather small in the assemblages under investigation (Gehlen 2009: 407-413) they might have served better for the production of flakes than of blades. In this way the high loadings of cores without modifications as well as of tools made of flakes on the second factor can be explained. This suggests a local production, which is further corroborated by the positive correlation between a high amount of artefacts with cortex and the second factor.

It is quite noteworthy that both factors are related in some way with flint production. The second factor though only seems to display the degree of a specific local production, which is not connected to high production rates in general, because the variable "flakes without modifications" has no meaningful loading on this factor. As a possible explanation this local production might have mainly been associated with the accommodation of local flint gravels.
This idea can be tested by investigating the relation of the second factor to the amount of gravel in the assemblages. In order to do this the factor scores are calculated. In principle, they represent a measure for each observation, meaning each flint assemblage on each factor (Hair *et al.* 2010: 139). Thereby every assemblage is provided with two additional variables, the factor scores for both factors. In a further step they will be used to position the observations in a diagram, in which the factors serve as axes. At this point a correlation between the factor score of the second factor and the percentage of gravel is calculated. The high correlation with a coefficient of 0.788 and a statistical significance of 99% confirms the idea that the second factor is connected to the local production of flint artefacts from gravel. It is therefore very probable that settlements with higher amounts of gravel flint also display a comparable high score on the second factor. Additionally this can be observed in Fig. 3.

Thus both factors could be interpreted in a satisfying way. In the next step, as already described above, the sites are displayed in a diagram (Fig. 4), whereby their position is defined by the factor scores. The factors function as axes. Differences between the sites in connection to the provision of flint are expected to become visible. Since the second factor displays probably mainly a local production of gravel flint, the position of the sites on the first factor is of much more interest.

Indeed, the flint assemblages scatter over a wide range on the first factor. According to the interpretation of the factor this is due to differences in production rates. Although a production of flint artefacts took place in every site under observation, five sites are characterised through very high production rates. Included in this group is the site of Langweiler 10, which already stood out among the older assemblages on account of its high amount of flint of the type Rullen. Considering however the possibility that the assemblage from this site and Langweiler 12 could be from the same settlement the outstanding position of Langweiler 10 must be interpreted with care. Langweiler 12 is placed quite close to the centre of the first factor and therefore displays an average production rate. The site of Langweiler 10 might represent a specific area inside a settlement, where flint-knappers disposed of their waste products. However since most of the flint material from this location is of the type Rullen, this might hint to the beginning of a specialisation in procurement and distribution of that flint type (Gehlen and Schön 2009a: 272). Most of the other sites with high production rates, Langweiler 1, Langweiler 5 and Langweiler 7, are located in close proximity to Langweiler 10 and

12. Hambach 59 is an exception. Apart from those the other sites display either an average production rate, lying close to the centre of the first factor, or a low production rate, which is represented by the positive pole of the factor.

From those results two theses can be deduced:

1) The differences in production rates on the sites indicate an exchange of flint materials between the settlements of the research area, whereby probably blades, flakes and cores were passed on from sites with high production rates to those with lower production rates. Since mainly the settlements of Langweiler are characterised by high production rates, they played an important role in the supply and distribution of flint (Gehlen and Schön 2009b: 605). Furthermore it is possible to assume, that in those settlements a tradition to procure flint of the type Rullen was established in early middle neolithic times and carried on throughout Rössen. Additionally, a distribution of flint artefacts was probably also conducted by the site of Hambach 59 (Gehlen and Schön 2009b: 605). However one curious fact yet remains unexplained in this scenario: Artefacts with cortex usually occur at an early stage of flint production during the preparation and the early reduction sequences (Gehlen 2009: 389; Zimmermann 1988: 646-647). Studies on early neolithic flint assemblages in the research area demonstrated that on sites with high production rates often an early reduction of cores took place , though not always (Zimmermann 2002: 31-32). In the middle neolithic assemblages nothing similar can be observed so far. On the contrary, the variable "artefacts with cortex" has no meaningful loading on the first factor. Its loading is even positive, suggesting that sites with average to low production rates may possibly have a higher percentage of artefacts with cortex than the others (Gehlen 2009: 389). This might be due to an additional supply with gravel flints on some of these sites. That a high amount of artefacts with cortex is connected to the local production of gravel displayed on the second factor is quite plausible, if assumed that there was no distribution of blades or flakes of this material. Most importantly, it remains unclear at this point if on sites with high production rates an early reduction of cores took place. Therefore an alternative thesis should be considered.

2) The result of the factor analysis might not refer mainly to an internal exchange of flint in the research area. It is possible that settlements with high production rates maintained external relationships with settlements outside the study area. In that case also settlements with average to low production rates might have been self-sufficient.

Both theses shall be tested in the following. To confirm the first thesis analyses of the reduction sequences in the flint assemblages of the research area are necessary. The second thesis requires the analysis of the interregional distribution of the flint that is mainly used in the research area, namely the Lanaye flint.

## 3.Analysis of reduction sequences

The first step to analyse differences in reduction sequences between the flint assemblages is to directly compare the percentages of artefacts with cortex. As displayed in Fig. 5 the percentages from sites with high production rates do not exceed the ones from other settlements (Gehlen and Schön 2009b: 593). In addition most of the assemblages from the early middle neolithic have a higher amount of artefacts with cortex than the ones from Rössen. This observation is no longer valid though, if only the artefacts with cortex from Lanaye flint are considered (Fig. 6). Therefore the higher amounts of cortex on the older sites are mainly due to an additional supply with gravel flint and other materials.

At this point it must be mentioned that from sites located in the study area and dating to the early neolithic (Linearbandkeramik) generally higher amounts of artefacts with cortex are observed than in the middle Neolithic. Taking this into account several authors suggested that during the Rössen period the first preparation of cores did not take place in the settlements but possibly at the extraction sites of the flint (Zimmermann 1995: 65; Gehlen 2009: 389). However there are differences between the reduction techniques from the early neolithic and the Rössen period: L. Fiedler (1979:75) observed a higher amount of core tablets in the older assemblages, while in Rössen crested blades are more numerous. Thus for the early neolithic a preparation of the striking platform is more common, while in Rössen the preparation of the reduction face was of higher importance (Fiedler 1979: 75; Zimmermann 1988: 662-663). Considering those differences it must be questioned if a direct comparison of the amount of artefacts with cortex between the time periods is appropriate.

Accordingly in this study a different approach for comparison is proposed: As an integral part of the documentation of the flint artefacts it was recorded to what amount the dorsal face and the platform remnant of blades and flakes are covered with natural surfaces, including cortex and joint planes. Three categories were distinguished: covered to less than 1/3, covered to more than 1/3 and covered completely (Zimmermann 1988: 583). In the following, the flint assemblages will be compared in regard to those categories. The artefact numbers of most of the assemblages on hand are not very high and in the following only artefacts with natural surfaces are considered. Therefore only three sites with a high number of artefacts were taken into account as well as one site from the early neolithic, namely Langweiler 8. The settlements dating to the middle neolithic are Aldenhoven 1, Inden 1 and Langweiler 1. While the sites of Aldenhoven 1 and Inden 1 are characterized by rather low production rates, for the site of Langweiler 1 a high degree of production was detected. The additional settlement of Langweiler 8 can also be classified as a site with a high production rate (Zimmermann 1988: 646-651). As shown in Fig. 7 most artefacts with complete covering occur in Langweiler 1. This is especially distinctive among the blades. The two sites of Inden 1 and Aldenhoven 1 display in all cases the lowest amount of artefacts completely covered with natural surfaces. Those observations indicate that in Langweiler 1 more often than in the other settlements the first or at least an early reduction of cores took place. This is also valid, though to a lesser degree in comparison to the early neolithic site of Langweiler 8. Therefore the idea that in Rössen cores were prepared at extraction sites might not necessarily be maintained. Additionally, the differences between Langweiler 1 and the two assemblages of Aldenhoven 1 and Inden 1 could not be detected by only comparing the percentages of artefacts with cortex in the flint assemblages (see Fig. 5). A possible explanation is that not only waste but also flint products, which were passed on between the settlements, were covered with cortex (Nowak 2013).

The next analytical step is dedicated to the comparison of artefact measures between the three settlements of Langweiler 1, Aldenhoven 1 and Inden 1, since their flint assemblages possess a sufficient number of artefacts. The analysis is based on the assumption that due to reduction not only the core decreases in size but also the proportions of the artefacts produced from that core get smaller (Zimmermann 1995: 84-89). Therefore at sites with an early reduction of cores artefact proportions may exceed those from other sites. Additionally, tools might have been in use for a shorter time in settlements with a better supply. Consequently they would be longer and possibly wider at the time of their discard in comparison to tools from other sites (Zimmermann 1999: 157).

In Fig. 8 the proportions of tools made of blades and blades without modifications are shown. The best distinction can be made between the lengths of the artefacts from the three sites. In all three assemblages the length of the tools made of blades surpasses those of the blades without modifications. This suggests that the longest blades were usually modified to tools (Gehlen and Schön 2009b: 541). Although in Inden 1 the difference between the two categories of blades is less striking. The longest artefacts are in both cases from the site of Langweiler 1, which further corroborates the idea that there cores were reduced at an early stage. A U-Test after Mann-Whitney was preformed to evaluate the statistical significance of the differences between the artefact lengths. This test can only be used on two datasets at a time (Hirsig 2001: 5.39-5.45). The results are displayed in Table 2.

The lengths of tools made of blades from Langweiler 1 differ significantly from those of the other assemblages. It is quite of interest however that there is no significant difference between the length of blades without modifications from Langweiler 1 and Inden 1, while both are significantly longer than the ones from Aldenhoven 1. This might be explained by taking into account the result of the factor analysis. It can be observed in Fig. 4 that Aldenhoven 1, though located on the positive range of the first factor, is still characterised be a higher production rate than Inden 1. Probably the settlers of Aldenhoven 1 received cores, which were already reduced to a certain degree, and produced their blades mainly from those. Inden 1 on the other hand received more often blades from sites with high production rates that kept the long ones and distributed blades of lesser sizes. Those were usually still longer than the blades produced in Aldenhoven 1.

In this part of the study it was shown, though mainly taking in to account three assemblages from the Rössen period that on the site of Langweiler 1 a high production rate correlates to an early reduction of cores. Therefore it is very likely that the settlers from Langweiler 1 supplied not only themselves with flint, but also distributed cores as well as blades and maybe flakes to other settlements in the research area. Hence, the comparisons of reduction sequences between three Rössen assemblages support the first thesis, which was proposed to explain the result of the factor analysis. Furthermore, this implicates that settlements like Aldenhoven 1 and Inden 1 were not self-sufficient, as suggested in the second thesis. Nonetheless in the next step of this study the second thesis shall be tested in regard to a possible distribution of flint beyond the research area.

## 4.Interregional relations

According to the second thesis some settlements in the study area display higher production rates due to external exchange relations. In order to test this additional data was collected from several publications (Table 3). Since the flint of the types Rullen and Rijckholt is the major raw material used in the research area, its distribution on an interregional scale is investigated. On this scale the distinction between the two types is less important and shall not be made.

In the following the percentage of Lanaye flint from different sites was interpolated and visualized by isolines (Fig. 9). Since the distribution patterns in the early middle neolithic and the Rössen period are quite similar (Nowak 2013), in this study only one analysis was preformed taking into account both time periods. The distribution map mainly represents parts of south and middle Germany. Assemblages from Lorraine were also considered dating to the Rössen period (Blouet et al. 2007). Additionally sites were added, on which Lanaye flint occurs though no percentages were available, as for example from two sites in Luxembourg (Le Brun-Ricalens and Valotteau 2007). Since the study area is not located between those sites and the flint outcrops, it is likely that the flint occurring in Luxembourg was not passed on from the lower Rhineland.

Compared to studies on the distribution of flint in the early neolithic (Zimmermann 1995) the data from the middle neolithic is not very extensive. The dotted isolines on the distribution map pass through regions were no data could be collected. Unfortunately this affects important regions like Hesse. Thus an interpretation must be preformed with care.

On the whole, no high amounts of Lanaye flint can be observed outside of the research area. Sites to the east in Westphalia still display considerable percentages. As mentioned above, data from Hesse is missing, except for two sites in the south [L139, 82]. Beyond those regions the Lanaye flint did not play an important role and occurs only in very small amounts. This was quite different in the early neolithic. During this time period the type Rijckholt occurred on a couple of sites in Franconia to more than 5% (Scharl 2010: 59) and in settlements from the Neckar mouth to about 10% (Lindig 2002: 125-127). While the middle neolithic assemblages from Westphalia consist to about 15 to 20% of Lanaye flint, most early neolithic assemblages from this region display usually far more than 30% of this flint (Gabriel 1974; Zimmermann 1995: 111-113).

Taking into account the fast drop of the percentage of Lanaye flint beyond the study area it is likely that settlements from other regions did not directly procure this flint. Therefore, as proposed in the second thesis, it is probable that Lanaye flint was passed on from the research area to other regions in the east and maybe the south. The distribution pattern can best be explained by a down-the-line exchange. Thereby settlements beyond a zone of direct access to the raw material sources passed on flint to neighbouring sites. Those in turn gave part of the flint they received to other settlements in their vicinity (Renfrew 1972: 465-466). The database for the middle neolithic is not very sufficient, though, and it is not possible to estimate the total amounts of the flint that was passed on from the study area. However the distribution of Lanaye flint in the middle neolithic is less extensive in comparison to the early neolithic. Accordingly it is likely that the high production rates in some sites of the research area are mainly due to an internal flint exchange, as proposed in the first thesis, and that an additional distribution to sites outside the research area took place.

## Conclusions

Through the application of a factor analysis different rates of production could be detected in the middle neolithic assemblages. Two theses were proposed to explain this result. According to the first thesis an exchange of flint material between the sites in the study area took place, while in the second thesis a distribution of flint outside of the lower Rhineland by the settlements with high production rates was considered. The results of further analyses support both theses. However, the first thesis explains the results of the factor analysis probably to a higher degree. According to this thesis some settlements within the research area were responsible for the procurement of Lanaye flint, mostly the type Rullen, and the distribution of cores, blades and probably flakes to other sites.

Acknowledgements : I would like to thank B. Gehlen and W. Schön for providing the data on most of the flint assemblages.

## References

Albers, H. J. and Felder, W. M., 1999. Die neolithische Abbautechnik vom Typ Aubel auf der Hochfläche der Limburger Kreidetafel als Konsequenz der postoligozänen Bildung einer Feuerstein-Residuallagerstätte. In Weisgerber, G., Slotta, R. and Weiner, J. (eds.), *5000 Jahre Feuersteinbergbau. Die Suche nach dem Stahl der Steinzeit*[3]. Bochum: Deutsches Bergbau-Museum Bochum, 67-79.

Allard, P., 2005. *L'industrie lithique des populations rubanées du Nord-Est de la France et de la Belgique*. Rahden/Westf.: Verlag Marie Leidorf (International Archaeology 86).

Blouet, V., Bour, J.-Y., Ganard, V., Klag, T., Lafitte, J.-D., and Petitdidier, M.-P., 2007. Le Néolithique moyen de la moyenne Moselle fran¸aise. In F. Brun-Ricalens, F. Valotteau and A. Hauzeur (eds.), *Relations interrégionales au Néolithique entre Bassin parisien et Bassin rhénan. Actes du 26e colloque interrégional sur le Néolithique Luxembourg, 8 et 9 novembre 2003*, Archaeologia Mosellana 7, 319-341.

Le Brun-Ricalens, F. and Valotteau, F., 2007. Le Néolithique moyen luxembourgeois: regards croisés entre le Bassin rhénan et le Bassin parisien. In: F. Brun-Ricalens, F. Valotteau and A. Hauzeur (eds.), *Relations interrégionales au Néolithique entre Bassin parisien et Bassin rhénan. Actes du 26e colloque interrégional sur le Néolithique Luxembourg, 8 et 9 novembre 2003*, Archaeologia Mosellana 7, 297-317.

Fiedler, L., 1979. *Formen und Techniken neolithischer Steingeräte aus dem Rheinland. Beiträge zur Urgeschichte des Rheinlandes III.* Köln: Rheinland-Verlag (Rheinische Ausgrabungen 19), 53-190.

Gabriel, I., 1974. Zum Rohmaterial der Silex-Artefakte im Neolithikum Westfalens und Nordhessens. In Alföldi, A. and Tackenberg, K. (eds.), *Kurt Tackenberg zum 75. Geburtstag*. Bonn: Rudolf Habelt Verlag (Antiquas 2/10), 25-45.

Gehlen, B., 2009. Silexinventare der Rössener Kultur im Rheinischen Braunkohlerevier. In: Zimmermann, A. (ed.), *Studien zum Alt- und Mittelneolithikum im Rheinischen Braunkohlerevier.* Rahden/Westf.: Verlag Marie Leidorf (Kölner Studien zur Prähistorischen Archäologie 1), 375-478.

Gehlen, B. and Schön, W., 2007. Céramique Linéaire récent – début du Néolithique moyen-Rössen dans le Bassin à lignite rhénan: les pièces lithiques comme reflets d'un monde changeant. In F. Brun-Ricalens, F. Valotteau and A. Hauzeur (eds.), *Relations interrégionales au Néolithique entre Bassin parisien et Bassin rhénan. Actes du 26e colloque interrégional sur le Néolithique Luxembourg, 8 et 9 novembre 2003*, Archaeologia Mosellana 7, 625-654.

Gehlen, B. and Schön, W., 2009a. Das frühe Mittelneolithikum: Steinartefaktinventare aus Langweiler 10, Hambach 260 und Langweiler 12. In: A. Zimmermann (Hrsg.), *Studien zum Alt- und Mittelneolithikum im Rheinischen Braunkohlerevier.* Rahden/Westf.: Verlag Marie Leidorf (Kölner Studien zur Prähistorischen Archäologie 1), 237-286.

Gehlen, B. and Schön, W., 2009b. Jüngere Bandkeramik – Frühes Mittelneolithikum – Rössen im Rheinischen Braunkohlerevier: Steinartefakte als Spiegel einer sich verändernden Welt. In: A. Zimmermann (Hrsg.), *Studien zum Alt- und Mittelneolithikum im Rheinischen Braunkohlerevier.* Rahden/Westf: Verlag Marie Leidorf (Kölner Studien zur Prähistorischen Archäologie 1), 587-611.

De Grooth, M. E. Th., 2007. Flint: procurement and distribution strategies; technological aspects. In P. van de Velde (ed.), *Excavations at Geleen-Janskamperveld 1990/1991*, Analecta Praehistorica Leidensia 39, 143-171.

De Grooth, M. E. Th., 2011. Distinguishing upper cretaceous flint types exploited during the Neolithic in the region between Maastricht, Tongeren, Liège and Aachen. In J. Meurers-Balke and W. Schön (eds.), *Vergangene Zeiten... LIBER AMICORUM. Gedenkschrift für Jürgen Hoika*. Bonn: Deutsche Gesellschaft für Ur- und Frühgeschichte e. V. (Archäologische Berichte 22), 107-130.

Hair jr., J. F., Black, W. C., Babin, B. J. and Anderson, R. E., 2010. *Multivariate Data Analysis. A global Perspective*[5]. New Jersey: Prentice Hall.

Hirsig, R., 2001. *Statistische Methoden in den Sozialwissenschaften. Eine Einführung im Hinblick auf computergestützte Datenanalysen mit SPSS für Windows[3]*. Zürich: Seismo.

Lindig, S., 2002. *Das Früh- und Mittelneolithikum im Neckarmündungsgebiet*. Bonn: Verlag Dr. Rudolf Habelt (Universitätsforschung zur Prähistorischen Archäologie 85).

Löhr, H., Zimmermann, A. and Hahn, J., 1977. Feuersteinartfakte. In Kuper, R., Löhr, H. and Lüning, J. (eds.), *Der bandkeramische Siedlungsplatz Langweiler 9. Beiträge zur neolithischen Besiedlung der Aldenhovener Platte II.* Bonn: Rheinland-Verlag (Rheinische Ausgrabungen 18), 131-266.

Lönne, P., 2000. *Das Mittelneolithikum im südlichen Niedersachsen. Untersuchungen zum Kulturkomplex Großgartach, Planig-Friedberg, Rössen*. Göttingen: Verlag Marie Leidorf.

Neubauer-Saurer, D., 1995. *Rohstoffversorgung und Silextechnologie im Frühneolithikum Südwestdeutschlands am Beispiel der Siedlungen Hilzingen, Scharmenseewadel, Grießen und Mühlhausen. Teil 1: Rohstoffe und Primärproduktion*. Freiburg i. Br.: Wissenschaft & Öffentlichkeit (Freiburger Dissertationen 1).

Nowak, K., 2013. *Mittelneolithische Silex-austauschsysteme auf der Aldenhovener Platte und ihrer Umgebung*. Köln: unpublished phD.-thesis.

Peters, R., 2011. *Demographisch-kulturelle Zyklen im Neolithikum. Die Bandkeramik im Rheinland und die Phyner Kultur am Bodensee*. Köln: unpublished Magister-thesis 2011.

Renfrew, C., 1972. *The emergence of civilisation. The Cyclades and the Aegean in the third millennium B.C.* London: Methuen and Co.

Scharl, S., 2010. *Versorgungsstrategien und Tauschnetzwerke im Alt- und Mittelneolithikum. Die Silexversorgung im westlichen Franken*. Rahden/Westf.: Verlag Marie Leidorf (Berliner Archäologische Forschungen 7).

Schwitalla, G., 1997. Das Steinmaterial der alt- und mittelneolithischen Siedlungsplätze Hasselsweiler 1 und 2, Gem. Titz, Kr. Düren. In Lüning, J. (ed.), *Studien zur Besiedlung der Aldenhovener Platte und ihrer Umgebung. Beiträge zur neolithischen Besiedlung der Aldenhovener Platte VI.* Köln: Rheinland-Verlag GmbH (Rheinische Ausgrabungen 43), 322-382.

Spatz, H., 1991. *Der Langweiler Typus'. Ein Nachweis der Gruppe Blicquy im Rheinland*. Germania 69, 155-162.

Weiner, J., 1997. Die Maasschotter der niederrheinischen Bucht als Feuersteinlieferant für die bandkeramischen Siedlungen Langweiler 8 und 9. In Lüning, J. (ed.), *Studien zur Besiedlung der Aldenhovener Platte und ihrer Umgebung. Beiträge zur neolithischen Besiedlung der Aldenhovener Platte VI.* Köln: Rheinland-Verlag GmbH (Rheinische Ausgrabungen 43), 599-646.

Zimmermann, A., 1988. Steine. In Boelicke, U., von Brandt, D., Lüning, J., Stehli, P. and Zimmermann, A., *Der bandkeramische Siedlungsplatz Langweiler 8, Gemeinde Aldenhoven, Kreis Düren. Beiträge zur neolithischen Besiedlung der Aldenhovener Platte III.* Köln: Rheinland-Verlag GmbH (Rheinische Ausgrabungen 28), 569-787.

Zimmermann, A., 1995. *Austauschsysteme von Silexartefakten in der Bandkeramik Mitteleuropas*. Bonn: Verlag Dr. Rudolf Habelt (Universitätsforschung zur Prähistorischen Archäologie 26).

Zimmermann, A., 1999. Neolithische Steinartefakte – ihre Merkmale und Aussagemöglichkeiten. In Preuß, J. (ed.), *Das Neolithikum in Mitteleuropa. Kulturen – Wirtschaft – Umwelt vom 6. bis 3. Jahrtausend v. u. Z. Übersichten zum Stand der Forschung*. Weissbach: Beier&Beran, 137-158.

Zimmermann, A., 2002. *Landschaftsarchäologie I: Die Bandkeramik auf der Aldenhovener Platte*. Berichte der Römisch-Germanischen Zentralkommission 83, 2002, 17-38.

Zimmermann, A., Meurers-Balke J. and Kalis, A., 2006. Das Neolithikum. In Kunow, J. and Wegner, H.-H. (eds.), *Urgeschichte im Rheinland. Jahrbuch 2005 des Rheinischen Vereins für Denkmalpflege und Landschaftsschutz*. Köln: Verlag des Rheinischen Vereins für Denkmalpflege und Landschaftsschutz, 159-202.

| variables | factor loadings factor 1 | factor 2 |
|---|---|---|
| tools made of flakes (%) | 0,653 | 0,56 |
| tools made of blades (%) | 0,891 | 0,07 |
| flakes without modifications (%) | 0,891 | -0,281 |
| cores without modifications (%) | 0,055 | 0,891 |
| artefacts with cortex (%) | 0,254 | 0,516 |

Table 1 : The loadings of each variable on both factors.

| **Length of tools made of blades:** | |
|---|---|
| Langweiler 1 – Inden 1 | p=0.000 |
| Langweiler 1 – Aldenhoven 1 | p=0.000 |
| Inden 1 – Aldenhoven 1 not significant | (p=0.494) |
| **Length of blades without modifications**: | |
| Langweiler 1 – Inden 1 not significant | (p=0.771) |
| Langweiler 1 – Aldenhoven 1 | p=0.000 |
| Inden 1 – Aldenhoven 1 | p=0.000 |

Table 2 : Statistical significance of differences between artefact length. The test results of the U-Test after Mann-Whitney.

**research area**

| no. | site | location |
|---|---|---|
| Ha59 | Hambach 59 | Hambach |
| In1 | Inden 1 | Aldenhovener Platte |
| In3 | Inden 3 | Aldenhovener Platte |
| Ald1 | Aldenhoven 1 | Aldenhovener Platte |
| Lw1 | Langweiler 1 | Aldenhovener Platte |
| Lw5 | Langweiler 5 | Aldenhovener Platte |
| Lw7 | Langweiler 7 | Aldenhovener Platte |
| Ha471 | Hambach 471 | Hambach |
| Nm1b | Nidermerz 1b | Aldenhovener Platte |
| Has1 | Hasselsweiler 1 | Hasselsweiler |
| Ha260 | Hambach 260 | Hambach |
| Lw10 | Langweiler 10 | Aldenhovener Platte |
| Lw12 | Langweiler 12 | Aldenhovener Platte |

**sites without percentage, but where Lanaye flint occurs**

| no. | site | references |
|---|---|---|
| H447 | Holzhausen | Gabriel 1974: 25-26 |
| EJ | Jouy aux-Arches | Blouet et al. 2007: 326 |
| Die | Diekirch | Le Brun-Ricalens and Va lotteau 2007: 303 |
| WK | Waldbillig | Le Brun-Ricalens and Va lotteau 2007: 304 |
| Mh | Müddersheim | Gehlen and Schön 2009b: 591 |

**sites outside of the research area**

| no. | site | references |
|---|---|---|
| F5 | Flévy «Massey Ferguson» | Blouet et al. 2007: 324-326 |
| U2 | Uckange «Cité des Sports» | Blouet et al. 2007: 327-328 |
| W104 | Werl | Gabriel 1974: 25-26 |
| B62a | Bochum-Laer | Gabriel 1974: 25-26 |
| D89 | Deiringen/Ruloph | Gabriel 1974: 25-26 |
| W111 | Werl | Gabriel 1974: 25-26 |
| G440 | Gudenberg | Gabriel 1974: 25-26 |
| M465 | Maden 465 | Gabriel 1974: 25-26 |
| M460 | Maden] 460 | Gabriel 1974: 25-26 |
| B64 | Bochum-Werne | Gabriel 1974: 25-26 |
| W110 | Werl | Gabriel 1974: 25-26 |
| N2 | Ladenburg | Lindig 2002: 127; 223 |
| N41 | Wallstadt | Lindig 2002: 127; 234-235 |
| Gr14 | Großenrode-14 | Lönne 2000: 101-103. |
| Mü | Mühlhausen | Neubauer-Saurer 1995: 26-27 |
| F13 | Burgbernheim | Scharl 2010: 220 |
| F15 | Euerfeld-Rothof | Scharl 2010: 220 |
| F17 | Hemmersheim | Scharl 2010: 220 |
| F22 | Schwanfeld | Scharl 2010: 220 |
| F14 | Dettelbach-Neusetz | Scharl 2010: 220 |
| F19 | Ippesheim | Scharl 2010: 220 |
| F20 | Rieden | Scharl 2010: 220 |
| F21 | Schwanfeld | Scharl 2010: 220 |
| F23 | Seenheim | Scharl 2010: 220 |
| 82 | Überau | Zimmermann 1995: 17 |
| L139 | Trebur-Tannböhl | Zimmermann 1995: 20-21 |

Table 3 : The database for the distribution map.

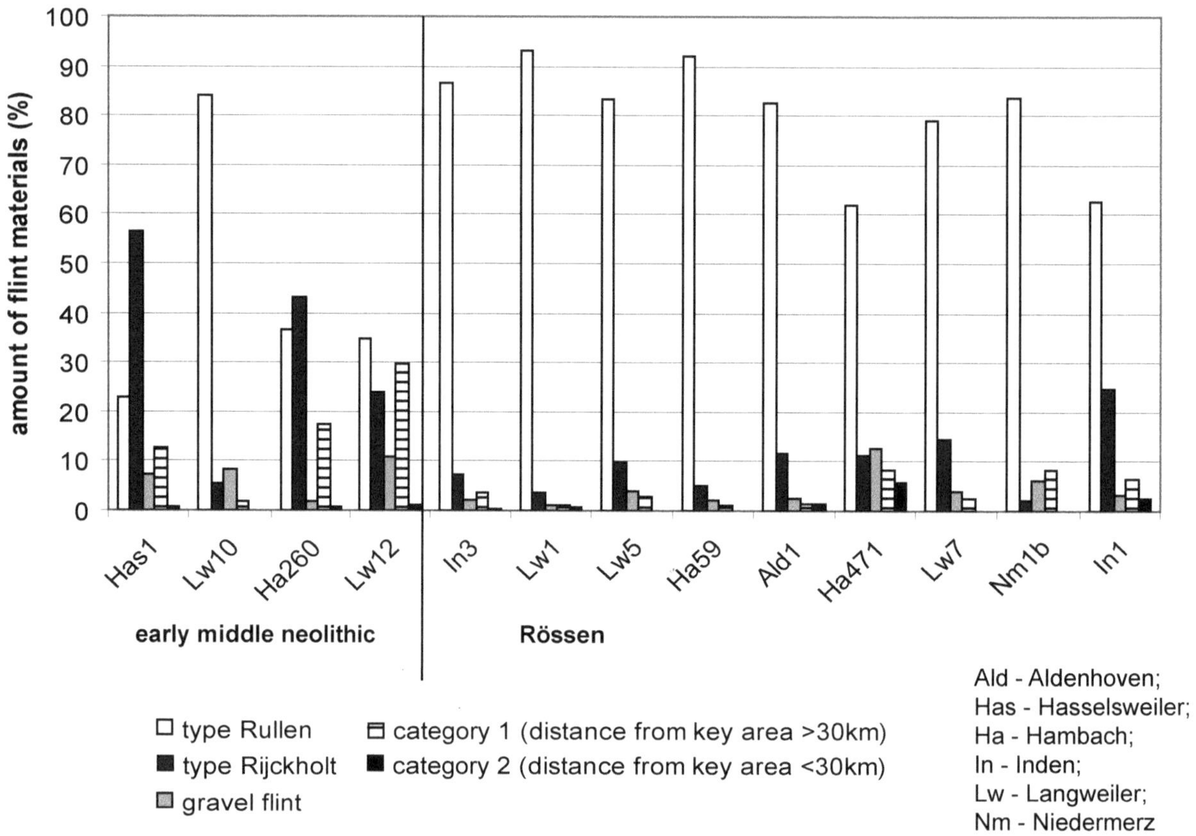


Fig. 1 : The raw materials in the middle neolithic flint assemblages.

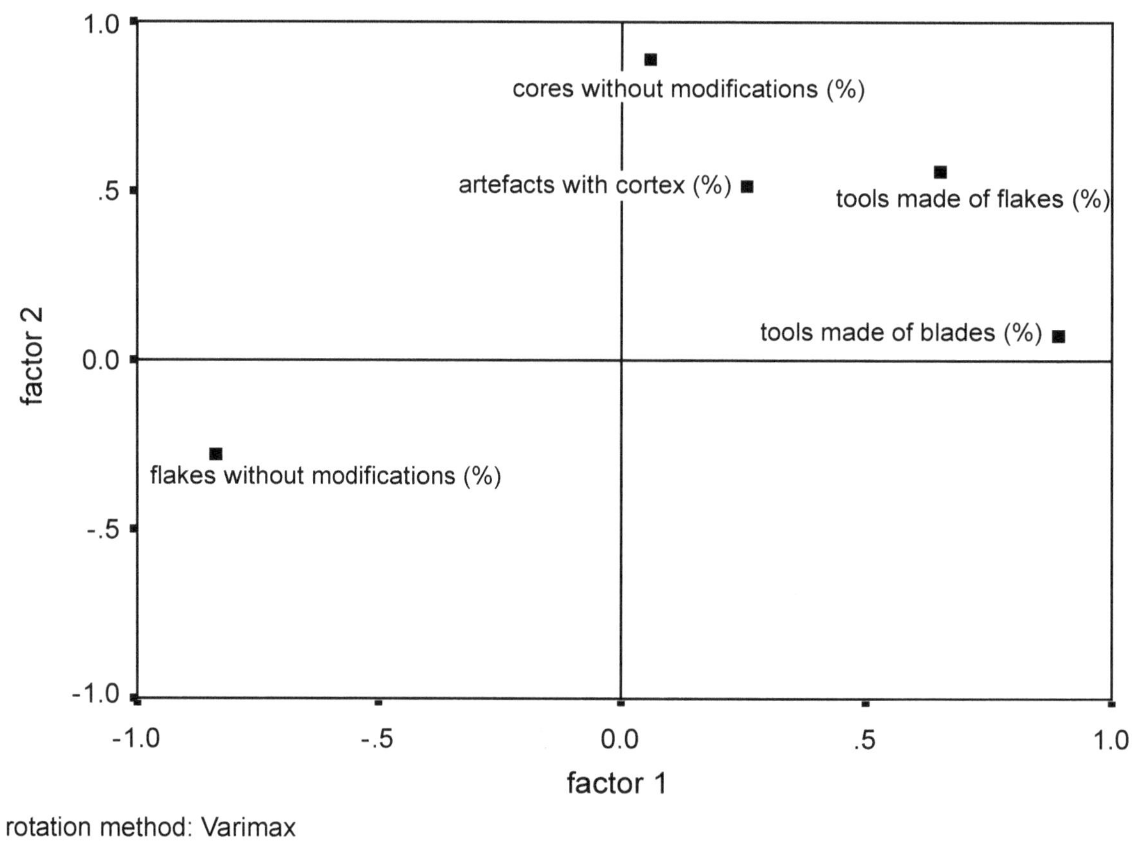


Fig. 2 : Diagram of the factor loadings.

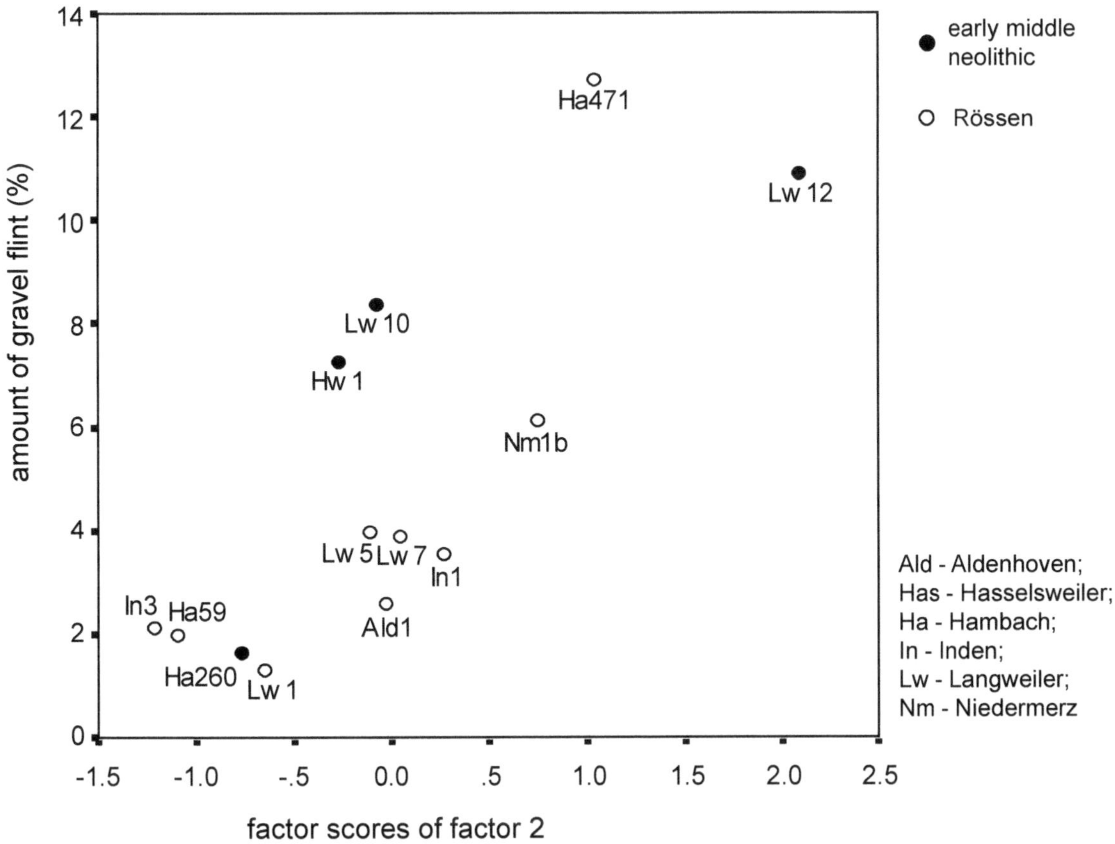


Fig. 3 : Diagram of the factor scores of the second factor and the percentage of gravel in the flint assemblages.

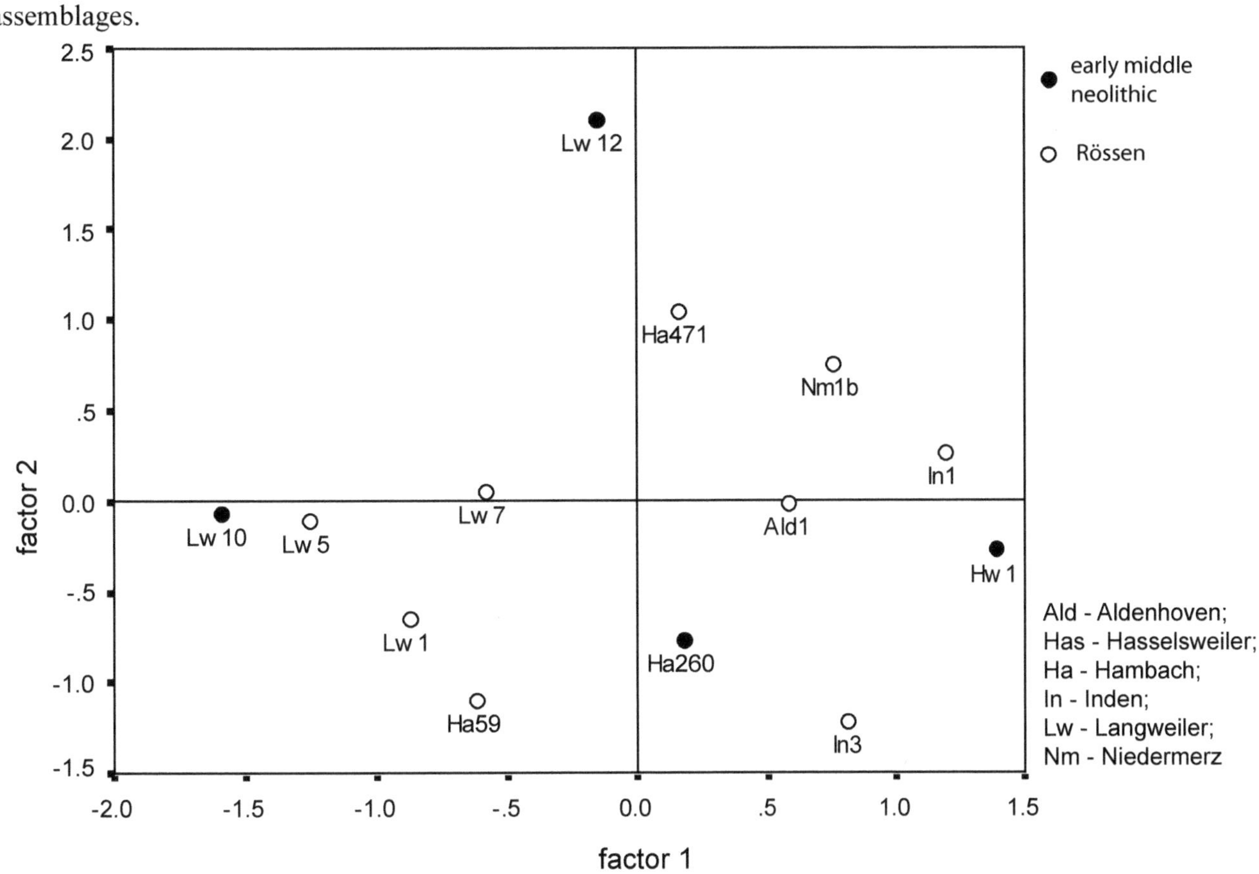


Fig. 4 : Diagram of the factor scores.

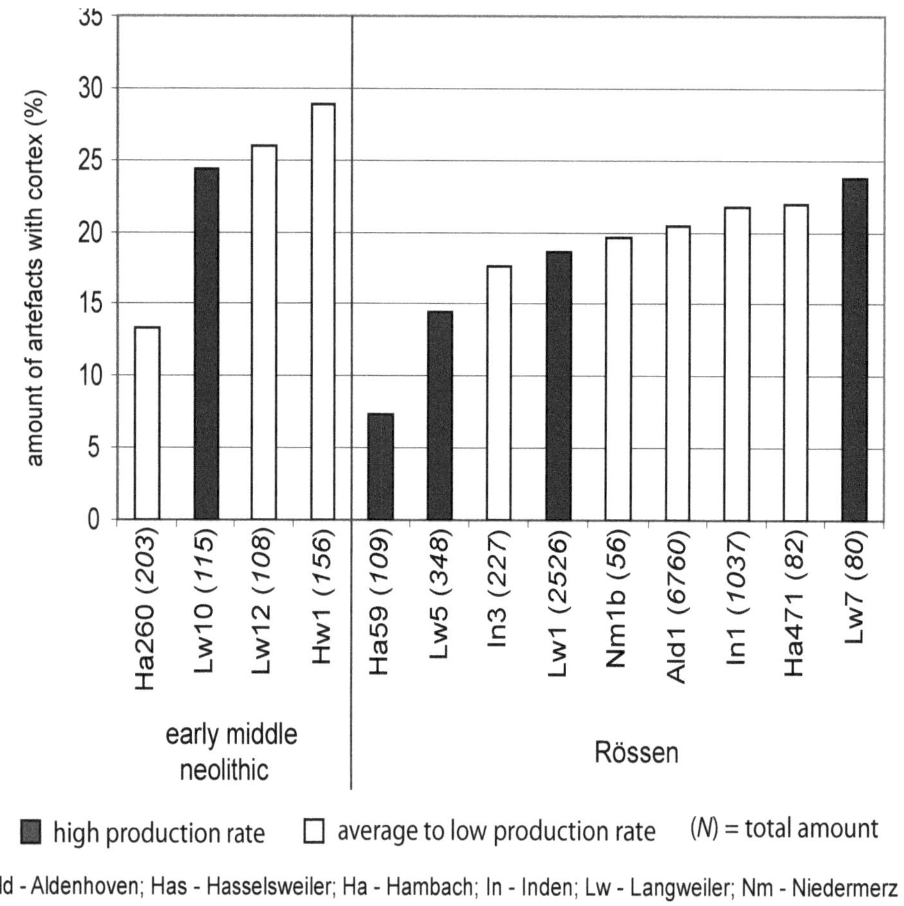


Fig. 5 : The percentages of artefacts with cortex.

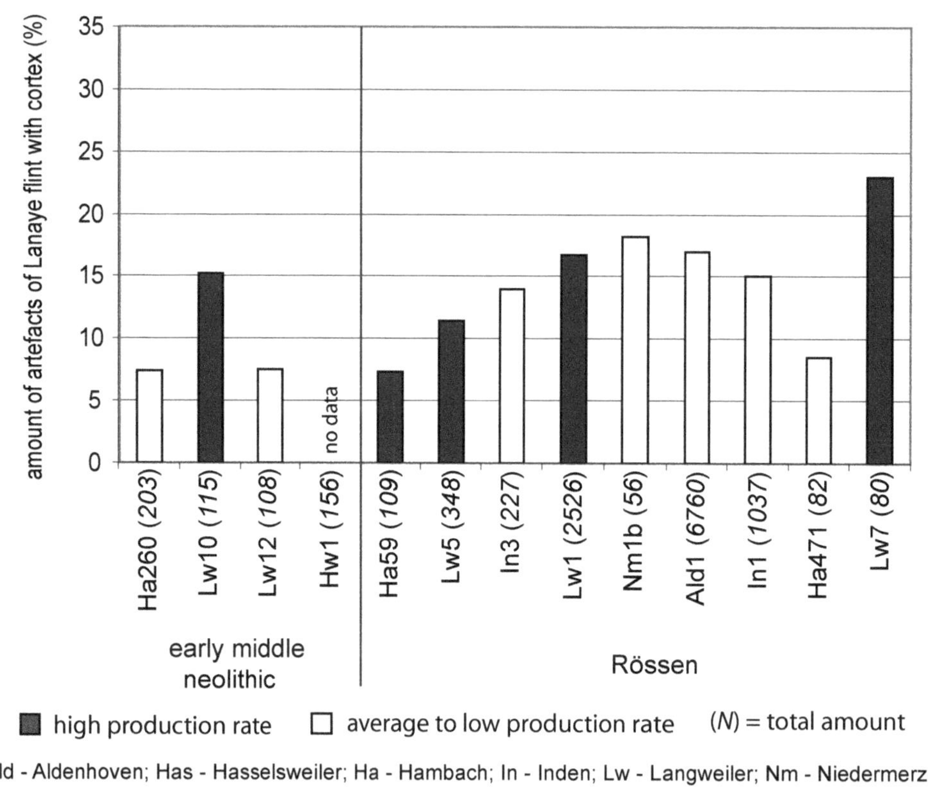


Fig. 6 : The percentages of the Lanaye flints with cortex.

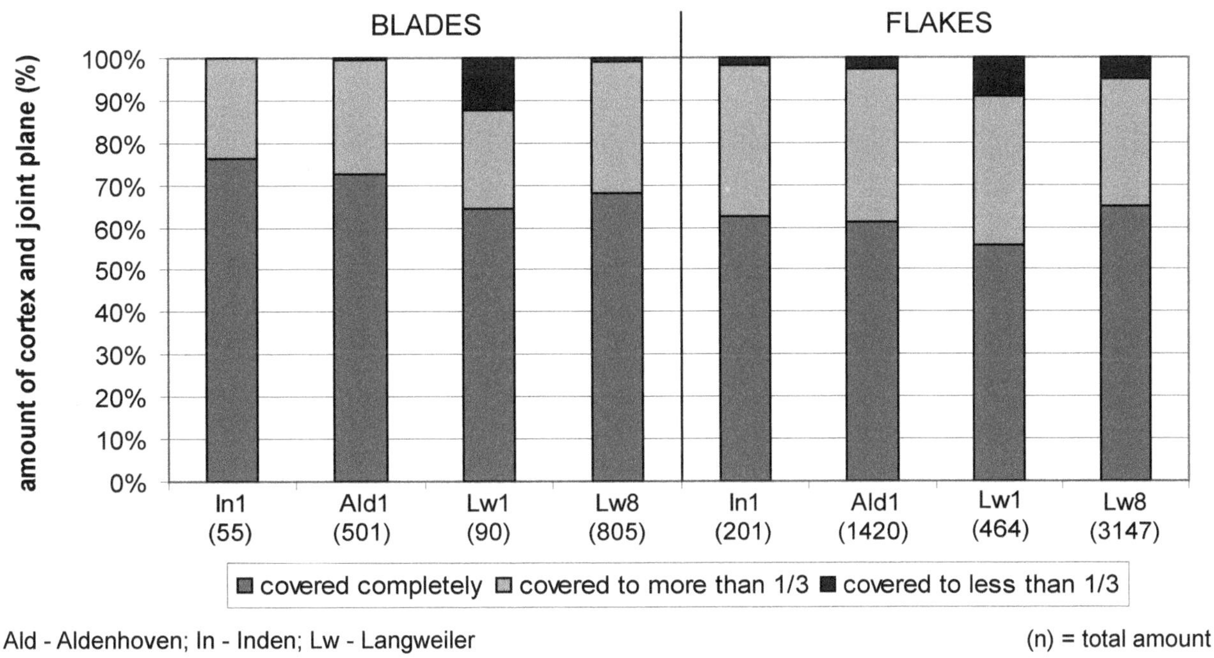


Fig. 7 : The coverage of blades and flakes with natural surfaces.

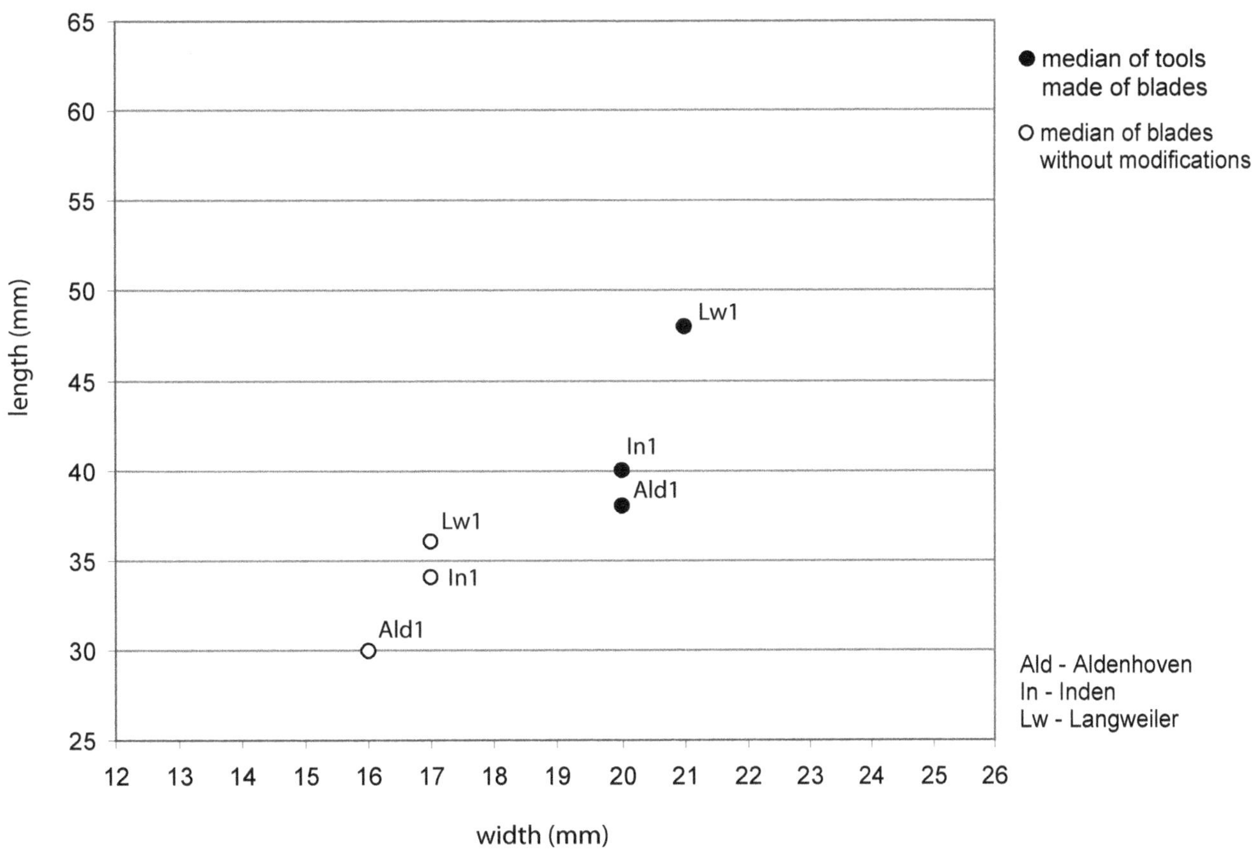


Fig. 8 : Medians of length and width of tools made of blades and blades without modifications.

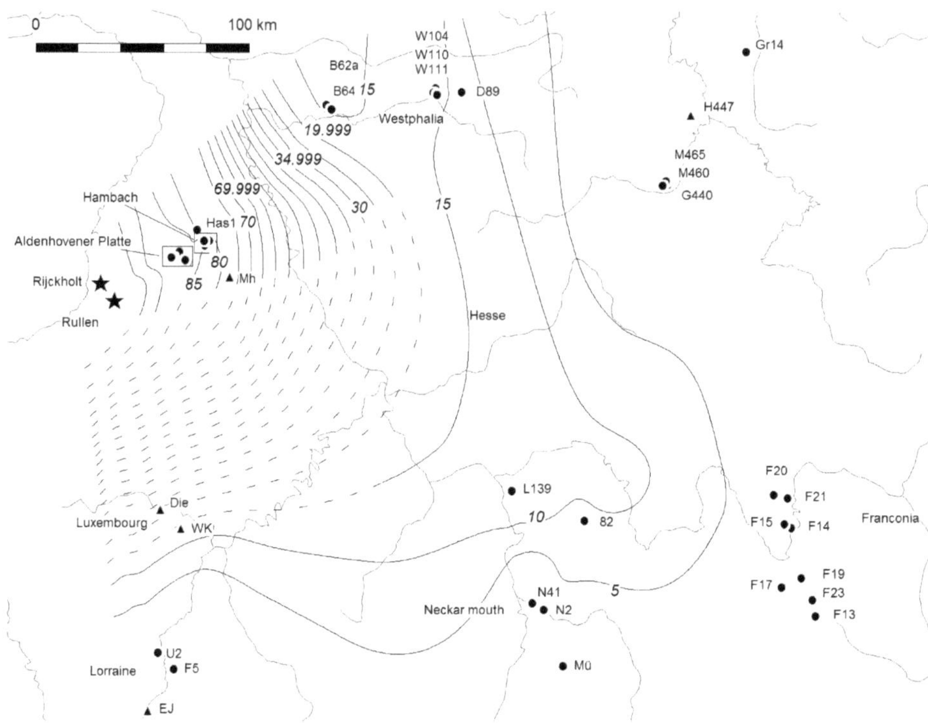


Fig. 9 : The Interpolation of the percentage of Lanaye flint during the middle neolithic (interpolation method = natural neighbour). Spots – sites included in the interpolation; triangles – sites without percentages, but were Lanaye flint occurs (not considered in the interpolation).

# Contemporary Rock Crystal Mining in Minas Gerais, Brazil – an Ethno-Archaeological Case Study

Michael BRANDL, Gerard TRNKA

## Abstract

A contemporaneous observation of mining techniques for crystal quartz in central-east Brazil provides a possible analogy to prehistoric mining in european contexts for the Neolithic.
The techniques of the miners - *garimpeiros* - are quite simple and the exploitation is undertaken without sophisticated social organisation and planning.

**Keywords :** ethnoarchaeology; mining; *garimpeiros;* crystal quartz; Brazil.

## Introduction

In 2011, the authors had the opportunity to examine contemporary, non-industrial rock crystal mining in Minas Gerais, Brazil, after the XVI th World Congress of the International Union of Prehistoric and Protohistoric Sciences (IUPPS/UISPP) in Florianópolis (State of Santa Catarina, Brazil) 4-10.9.2011.
This provided invaluable insight into individual mining activities conducted with methods and under circumstances comparable to prehistoric conditions. The organisation and techniques of Neolithic mining are elusive research issues due to the limited archaeological evidence beyond the primary remnant evidence such as mining tools and mining traces. Especially difficult undertakings are interpretations of prehistoric shaft mining. Numerous studies constructed theories and theoretic models as to how prehistoric societies were able to produce extensive and at the first glance sophisticated mining areas such as Arnhofen in Lower Bavaria or Grimes Graves in Great Britain. The results of our ethno-archaeological investigation in Brazil are suitable to challenge some of the popular ideas concerning prehistoric deep-shaft mining and the organization of mining communities.

## The Cabaçaco mining area

Cabaçaco or Cabassaco is located in the Jequitinhonha valley ca. 20 km north-east of Diamantina, an old diamond and gold mining town in central Minas Gerais, central-eastern Brazil. The mining area is situated in a desert environment alongside the foothills and partly on the slopes of the Espinhaço Mountain range, which extends over 1200 km from the region around Belo Horizonte to the northern border of the Bahia state. Its width ranges between only few to over 100 kilometers.

The Cabaçaco mining district extends in a north-western direction alongside the Jequitinhonha Valley, forming an approximately 2 km long "belt" running from south-east to north-west, with an approximate width of ca. 1 km (UNESCO 2013; Fig.1).

## Geological setting

Stoiber et al. (1945) conducted research in order to explore the relevance of quartz from Brazil for military supplies (i.e. quartz oscillator plates for radio frequency control) of the US army during WW II. They established geological parameters valid for the quartz bearing zones throughout Brazil:

*The deposits of quartz crystal in Brazil appear to form a "mineral province". Features which characterize this province that appear to be significant may be summarized as:*
*1. The deposits lie in four elongated belts, widely separated and roughly parallel to one another.*
*2. Most of the deposits in bedrock are in Paleozoic sedimentary rocks which are not highly folded.*
*3. Most of the area in which the deposits occur is characterized by deep weathering.*
*4. Most of the deposits in bedrock are quartz veins, or flat-lying blanket deposits of milky quartz roughly parallel to the enclosing sediments, or a combination of these two types. The clear crystal usually occurs in vugs or pockets and appears to have formed later than the bulk of the milky quartz.*
*5. The deposits are characterized by very few minerals in important quantity other than milky quartz.*
*6. Veins are apparently unrelated to other mineralization.*
*7. The veins are exploited for no product other than quartz crystal.*

*8. With few exceptions single deposits produce not more than 500 pounds of usable crystal per month.* (Stoiber et al. 1945, 228).

The study area for this investigation is located in one of the aforementioned "elongated belts" in Minas Gerais, closely related to the Espinhaço Mountain Range. It displays the largest and most extensive Precambrian orogenic belt within the territory of Brazil, representing a watershed between the São Francisco River basin and river basins draining into the Altantic Sea (UNESCO 2013).
Geologically, Cabaçaco is located in the Espinhaço Supergroup. Main constituents are sandstones, conglomerates, pelites and quartzites (Battilani et al. 2007, 322).
The geological member the rock crystal deposits are associated wit is part of the Sopa-Brumadnho formation, comprised of fine-middle grained quartzite with admixtures of mica and layers of iron oxides. Locally restricted rock masses show monomictic meta-conglomerates (Fig. 1).

## Theoretical models concerning Neolithic mining

Some ideas concerning Neolithic mining activities need to be addressed in order to comprehend the value of our observations in Brazil. In addition to "conventional" raw material procurement – i.e. collecting and digging of shallow pits for extraction – the mining of lithic raw materials in the Neolithic was carried out by digging shafts, sometimes over 10 m in depth. This does not necessarily implicate a highly sophisticated knowledge, as we were able to experience first hand at Cabaçaco.

Roth (2008) undertook investigations concerning the history of economics at the Arnhofen mining site in Lower Bavaria, comparing available data from different Middle and Late Neolithic (4400 – ca. 2800 BC) chert and flint mining sites throughout Central and Western Europe. He found similar patterns regarding mining organization at an area extending between Central Poland and the Paris Basin, allowing for some general assessments.
It seems that areas potentially suitable for mining were divided amongst small groups, probably representing kinship groups. Such groups consisted of not more than 3 - 4 members, which can be deduced from the size of the shafts. The mining activities were carried out seasonally on a part time basis additional to the main activities that fall within the farming economy. Particular groups were in possession of use rights at certain areas, which were probably inherited over generations. Roth (2008) derives these assumptions from the spatial organisation of the mining area at Arnhofen. However, he found no indication for fully specialised miners, nor could he identify "industrial" elements for the mining organisation (Roth 2008, 911-932).

## Ethno-archaeological investigations 2011

We were able to document a range of features and activities providing insight into the community system, mining operation and social aspects of the *garimpeiros*, the miners at Cabaçaco. The following details were obtained in the course of personal communication and observation, certainly limited due to language problems and biased statements from the miners, however conducted as thorough as possible taking the circumstances into consideration (i.e. more or less random encounters and selectively presented areas and situations during a very limited timespan).

## Mining activities

*Deep (vertical) shafts*

The most striking features we observed were mining fields stretching over hundreds of square meters, intriguingly resembling Neolithic mining areas. Most of the deep shafts at Cabacaco are situated in the flat areas of the Jequitinhonha Valley, reaching a depth of 5 – 10 m. We were told that there existed shafts over 30 m deep located on the sides of the Espinhaco foothills, however, we were not able to verify this.

One mining shaft worked by the *garimpeiro* Raimondo Nonato dos Santos was accessed by Brandl first hand (12.9.2011). From the entrance to the bottom, it is between 1 – 1,5 m wide and ca. 8 meters deep. It declines steeply, following a quartz vein visible in the side walls. After approximately 3 m there is a step, ca. 1,5 m in length, this is followed by an adjacent 5 m deep lower shaft leading straight down to a small chamber in which two people can barely fit. There exist no protection systems such as pillars or planking to secure the loose sidewall material, which mainly consists of weathered sandstone and quartz sand. Descending the shaft is accomplished without ladder or climbing support, only using small steps crudely picked into the sidewalls and by using the shaft walls as support. The best way climbing down

such a shaft is in a slow spinning motion. The only illumination are two candles, skilfully handled during the risky descend by the miner, alternately placed on tiny ledges and/or craved steps at the sidewalls.
The small chamber at the bottom of the shaft is at the level of sandstone with cavities infilled with loose sediment (weathered quartz sand). The (now) sand-filled cavities are the "nurseries" of the rock crystals. They come in various sizes, from microscopically tiny to over one meter in length, predominantly depending on the cavity's dimensions.
The extraction of the rock crystals is a task which requires both, force and finesse. The hard sandstone is removed either manually using picks or hammer and chisel, whereas for larger rock parts explosives are employed (dynamite; Fig. 4). Once the rock crystal bearing level is reached, explosives are usually banned from the shafts (with some exceptions, as we were told by some miners, who stressed out the extreme danger they underwent due to falling rocks and possible instabilities of the surrounding material by re-accessing a shaft in which dynamite had been applied). Brandl was able to observe the procedure of extracting a medium – large, multi – peaked rock crystal specimen from the shaft he accessed together with the owner, Raimondo Nonato dos Santos. The latter had already prepared a cavity in the aforementioned sandstone layer. He chiselled away some rock material in order to open the hole up to approximately the size of a human's head. He then reached into the cavity using the entire length of his arm, grabbed the specimen and removed it carefully by shifting and turning it around, specifically avoiding damage to the fragile peaks of the crystals. If a crystal shows smallest traces of chipping or other damage, it is considered worthless (as we learned shortly before we left the miner's camp, when the *garimpeiros* poured buckets of minimally damaged crystals to our feet and insisted that we took more than we could carry, without accepting any reward). The miners stay in the shafts from 1 to 4 hours max, depending on the depth (oxygen regulation, indicated by the candles they carry) and the location. When a shaft's entrance is directly hit by the sun, they are forced to work only early in the morning and late in the evening.

Waste rock is removed applying a simple wooden tripod construction with buckets and ropes (Fig. 4). Typically, the discarded material is dumped directly outside and around the shaft's openings, creating the characteristic "mining field landscape" well known from prehistoric examples (Fig. 2 and Fig. 4). It seems that backfilling is not a common practice.

*Horizontal shafts (Mine tunnesl)*
We also encountered shafts that were horizontally tunnelled into the flank of a hill. The one we entered was worked by an older man around 70 (as he pointed out), sitting on a self-made wooden construction, with only one candle for illumination, hammer and chisel and a water bottle with attached hose to minimize the formation of fine dust (Fig. 3). The horizontal tunnel was ca. 1 m wide, 1,5 m high and carved out of the hard sandstone over 50 m in length, passing a natural chamber (which they called "El Bojo", the "bowl" or "bulge"), always following a massive quartz vein in the hope for a natural cleft containing large rock crystal specimens.

## Accidents

According to our guides, accidents happen repeatedly (which is not surprising considering the poor security standards employed at the deep shafts as well as for the horizontal mines). When individuals perish in mines, the corpses are not retrieved, the shaft is shut down and serves as their grave.

## Living conditions and social system

The miners or *garimpeiros* live under very rustic conditions without their families in simple huts and wooden sheds. According to our driver, the *garimpeiros* – exclusively men – are essentially forced to work at the mines because they belong to a low socio-economic class without other perspectives. Everybody hopes to find the "master load" helping them to escape poverty.
Although our possibilities of communication were extremely limited due to our ignorance of the Portuguese language, it came clear that the social system of the mining community is very flat and does not display any hierarchical relationship between the involved agents. Every *garimpeiro* asks the land owner – the Don – for permission to dig a shaft. After the permission is granted, he is free to decide where to set his "Claim" and dig a hole. Some of the observed shafts are fairly close to each other, however there are no bigger mining clusters run by kinship groups.
During the initial stage, small groups of miners work together in order to get through the considerably hard sandstone of the upper geological layers at Cabacaco, however when the shaft gets deeper and accordingly narrower only one person conducts mining. At the stage a shaft reaches the layer in which the rock crystal occurs, every *garimpeiro* is operator of his individual mine.

The miners keep the yield of their activities in their huts and sheds, and the specimens are typically placed on tables for display. They present their most impressive rock crystals with obvious proud. However, it is impossible to purchase selected pieces, because the ware is sold en gros to merchants who contract the *garimpeiros* with an adhesive system, sealing the merchandise and preventing the best pieces to be obtained by "treasure hunters".

## Conclusion

As a conclusion, the lacking of any social system in the miner's community was the most important observation at Cabaçaco. The *garimpeiros* are working under unimaginable circumstances, in a hostile environment and under constant danger. This can only be explained by the pure necessity of survival.

In fact, Brazilian rock crystal mining is a very simple operation: Whenever there is the time and opportunity to do so, the men are working at their shafts, alone or in small groups of three miners max. They stay at the mining area for a certain time period, and after digging their lot for selling to the distributers they leave to spend time with their families (in case they have got a family). Only a lucky finder is able to stay away from the mines after a season, most of the *garimpeiros* return whenever the money is spent and the life costs force them back into the desert.

It is of course dangerous – and at the same time tempting – to establish connections to Neolithic conditions. We will never be able to observe the background underlying Neolithic mining activities, however one thing becomes very clear: There is simply no need for a sophisticated social organisation and planning (let alone security systems) for creating a mining area that is at the first glance very impressive. We do not know the circumstances under which Neolithic mining took place, but we suggest the assumption that they were much more like the ones we observed at Cabaçaco than many elaborate theoretical models propose.

Appendix - The chert mining fields of Arnhofen-Abensberg (Southern Germany)

The Arnhofen and Abensberg region in Lower Bavaria is one of the most famous Neolithic mining areas in Europe (an overview is provided by Rind & Roth 2007; Rind 2009). In a bigger picture, the similarities between the rock crystal mines in Cabaçaco (Brazil) and the Arnhofen mining fields are evident – both sites illustrate narrow and deep mining shafts dug through weak sediments, both are lacking evidence of precaution and security systems, the shafts in working progress are situated close together etc. (Fig. 5). Research results indicate that during Neolithic times only small groups of miners worked at the same time at Arnhofen. Mining was conducted only seasonally and at not more than three shafts at once on average. This puts the estimated 20.000 shafts dating between the mid 6th and the 5th Millennium BC into perspective, reducing the "monumental" impression of labour at Arnhofen to a rational effort within Neolithic life.

## References


Rind, M.M. and Roth, G. 2007. Archäologie in Deutschlands größtem Feuersteinbergwerk, *Archäologie in Deutschland* 4, 2007, 8-13.

Rind, M.M. 2009. Zum Stand der Ausgrabungen in Deutschlands größtem Hornsteinbergwerk, Internationale Archäologie - *Studia honoraria 29*, 2009, 21-26.

Roth, G. 2008. *Geben und Nehmen. Eine wirtschaftshistorische Studie zum neolithischen Hornsteinbergbau von Abensberg-Arnhofen, Kr. Kelheim (Niederbayern)* [in IV Bänden], Dissertation zur Erlangung des Doktorgrades im Fach Ur- und Frühgeschichte an der Philosophischen Fakultät der Universität zu Köln , 2008. http://www.archaeologie-online.de/links/detail/19189.php

Stoiber, R.E., C. Tolmann and R.D. Butler 1945. Geology of quartz crystal deposits. *American Mineralogist* 30, 219-229.

Geological maps adapted from:
CODMIG (Companhia de Desenvolvimento Econômico de Minas Gerais)

http://www.codemig.com.br/site/content/parcerias/mapeamento_geologicoEf_espinhaco.asp?id=17E (query 2.3.2013)

UNESCO 2013: http://www.unesco.org/mabdb/br/brdir/directory/biores.asp?mode=all&code=BRA+06 (query 2.3.2013)

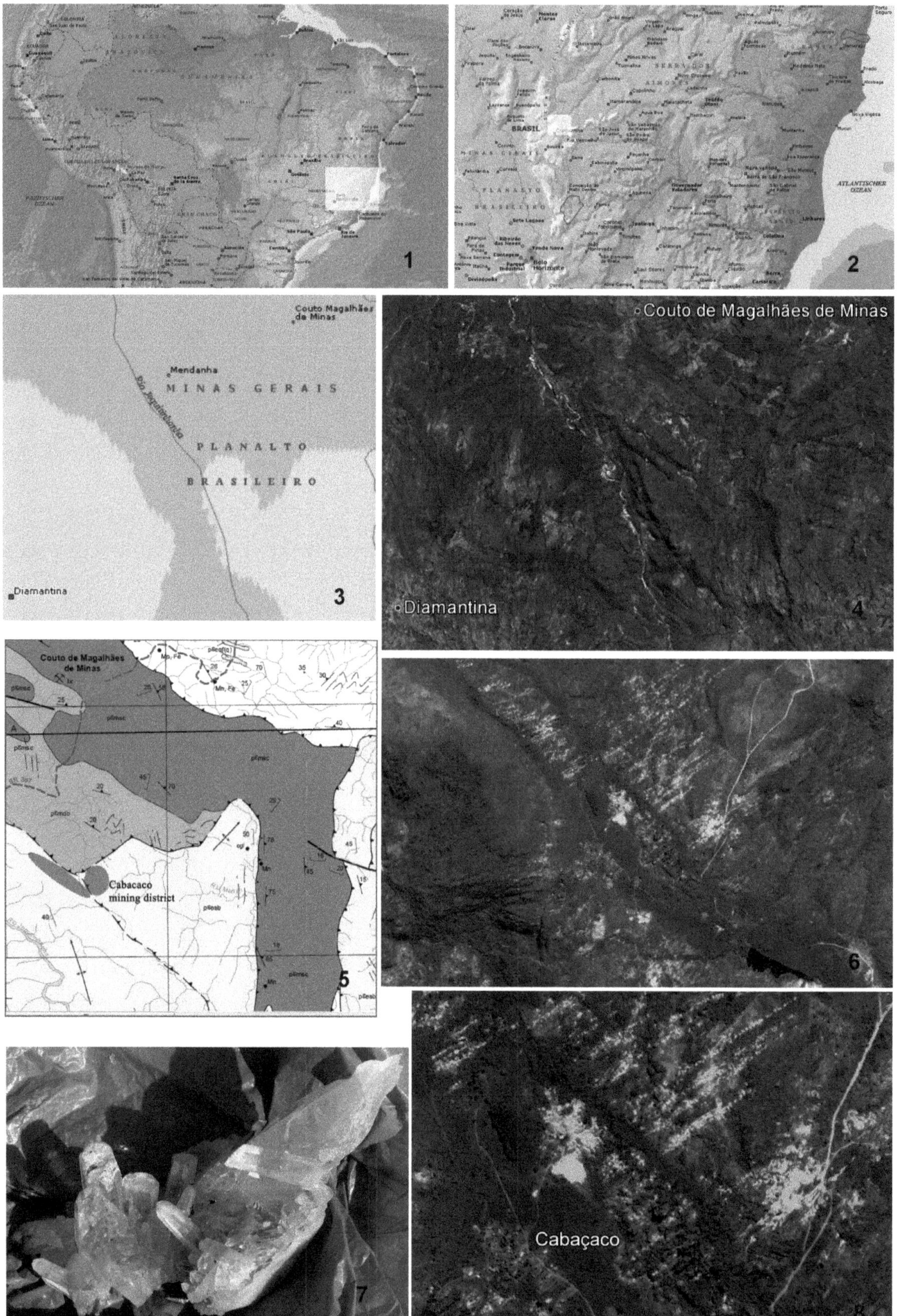


Fig. 1 : 1-2-3-4-6. Brazil. Minas Gerais. Geography of the region of Diamantina. 5. Geological map. 7. Sample of rock crystal; Cabaçaco. Average S 18.136148°, W 43.478578°, altitude 983 m. (12.9.2011).

Fig. 2 : Cabaçaco (Minas Gerais, Brazil). Village, mining field, shafts with quarz veins. 12.9.2011. Average S 18.132693°, W 43.480424°, altitude 989 m.

Fig. 3 : Cabaçaco (Minas Gerais, Brazil). Horizontal shaft (Mine tunnel). 12.9.2011.

Fig. 4 : Cabaçaco (Minas Gerais, Brazil). Mining field. 13.9.2011. Average S 18.134989°, W 43.473151°, altitude 1020 m.

Fig. 5 : Arnhofen (Lower Bavaria, Germany). Neolithic mining field. Excavations 2000- 2003. Fotos by permission of Michael M. Rind.

www.ingramcontent.com/pod-product-compliance
Lightning Source LLC
LaVergne TN
LVHW071219210926
848482LV00015B/855
*9781407312989*